YOUR BRAIN IS ABOUT TO DISCOVER:

—How each of the millions of neurons that comprise the brain operates both as an individual computer and in harmony with all others.

—Why you may stutter if you are left-handed.

—The symptoms, causes and treatment of brain disorders that can affect victims from the very young to the very old.

—The electricity that powers the brain.

—How language is learned and memory stored.

—The remarkable ability of the brain to control pain.

—The possible relationship between brain chemistry and mental disorder.

These samples give but a bare hint of the revelations that will expand your mind's knowledge of itself on every page of a book for both students and lay-people who seek to know as much as possible about the greatest single determinant of who we are and what we do.

INSIDE THE BRAIN

WILLIAM H. CALVIN, Ph.D., is a neurophysiologist in the Department of Neurological Surgery of the University of Washington School of Medicine. GEORGE A. OJEMANN, M.D., is on the faculty of the Department of Neurological Surgery, University of Washington School of Medicine. Both are the authors of numerous articles in professional journals.

MENTOR Books of Special Interest

INSIDE
THE BRAIN

MAPPING THE CORTEX, EXPLORING THE NEURON

William H. Calvin, Ph.D.,
and George A. Ojemann, M.D.

A MENTOR BOOK
NEW AMERICAN LIBRARY
TIMES MIRROR
New York and Scarborough, Ontario
The New English Library Limited, London

Copyright © 1980 by William H. Calvin and George A. Ojemann

Library of Congress Catalog Card Number: 80-81134

ACKNOWLEDGMENTS

Page ix: Quotation from "Little Gidding" in FOUR QUARTETS by T. S. Eliot copyright 1943 by T. S. Eliot: copyright 1971 by Esme Valerie Eliot. Reprinted by permission of Harcourt Brace Jovanovich, Inc., and Faber & Faber, Ltd. The opening figure is from the classic work of S. Ramon y Cajal.

Page 10: Figures opening Chapters 2, 3, 4, 5, 8, 10, 11, 14, 15, 16, and 17 are reprinted by permission of the publishers from THE POSTNATAL DEVELOPMENT OF THE HUMAN CEREBRAL CORTEX, Volumes I and VIII, by the late J. Leroy Conel, Cambridge, Mass.: Harvard University Press, copyright © 1939, 1967 by the President and Fellows of Harvard College, copyright © 1967 by Jesse Leroy Conel.

Page 15: Figure modified from that of Poritsky, Journal of Comparative Neurology, 135:447, reprinted by permission of the Wistar Press, copyright 1969.

Page 16: Figure modified from that of Fox in CORRELATIVE ANATOMY OF THE NERVOUS SYSTEM by E. C. Crosby et al, copyright 1962, by permission of McMillan Publishing Co., Inc.

Page 56: Figure reproduced from BRAIN MECHANISMS IN DIACHROME by W. J. S. Krieg, copyright 1955, with the permission of Professor Krieg.

Page 58: Quotation from THE BRAIN CHANGERS: SCIENTISTS AND THE NEW MIND CONTROL by Maya Pines, reproduced by permission of Harcourt Brace Jovanovich, Inc. Copyright 1973.

Page 60: Figure reproduced from "Differential changes with aging in old and new cortices," by M. E. Scheibel and A. B. Scheibel, in THE AGING BRAIN AND SENILE DEMENTIA edited by K. Nandy and I. Sherwin, Plenum Press copyright 1977, with the permission of Professor Scheibel and the Plenum Press.

(The following page constitutes an extension of this copyright page)

Page 65: Figure reproduced with the permission of Professor W. T. Greenough and *American Scientist* (Journal of Sigma Xi, the Scientific Research Society), copyright 1975 by Sigma Xi.

Page 68: Figures from "Elementary processes in selected thalamic and cortical subsystems—the structural substrates," by M. E. Scheibel and A. B. Scheibel, in THE NEUROSCIENCES, SECOND STUDY PROGRAM, edited by F. O. Schmitt, copyright 1970, with the permission of Professor Scheibel and the Rockefeller University Press.

Page 94: Figure modified from *International Journal of Neuroscience*, copyright 1973, with the permission of Professor Scheibel and of Gordon and Breach Science Publishers, Ltd.

Page 110: Figure modified from Walberg, *Experimental Neurology* 13:218, copyright 1965, with the permission of Professor F. Walberg and the Academic Press, Inc.

Pages 124, 126: Figures reproduced from the *Journal of Comparative Neurology*, copyright 1973 and 1977, with the permission of Professor Jennifer S. Lund and of the Wistar Press.

Page 138: This previously unpublished figure was provided by Professor Arne B. Scheibel, whom we thank for his many kindnesses.

Page 146: Figure reproduced from "Organization of crustacean neuropil. II. Distribution of synaptic contacts on identified motor neurons in lobster stomatogastric ganglion," by David G. King, in *Journal of Neurocytology* 5:239-266, copyright 1976, with the permission of Professor King and of Chapman and Hall Ltd.

Page 167: Figure reproduced from M. E. Scheibel et al, *Epilepsia* 15:55-80, copyright 1974 with the permission of the authors and of Raven Press.

Page 176: Excerpts from Lewis Thomas' "The technology of medicine," in *New England Journal of Medicine* 285: 1366-1368, December 9, 1971, with the permission of Professor Thomas and the publishers. This article was subsequently reprinted in *The Lives of a Cell*, Viking, 1974.

SIGNET, SIGNET CLASSICS, MENTOR, PLUME, MERIDIAN AND NAL Books are published in the United States by
The New American Library, Inc.,
1633 Broadway, New York, New York 10019
in Canada by The New American Library of Canada Limited,
81 Mack Avenue, Scarborough, Ontario M1L 1M8,
in the United Kingdom by The New English Library Limited,
Barnard's Inn, Holborn, London EC1N 2JR, England.

First Printing, September, 1980

1 2 3 4 5 6 7 8 9

PRINTED IN THE UNITED STATES OF AMERICA

CONTENTS

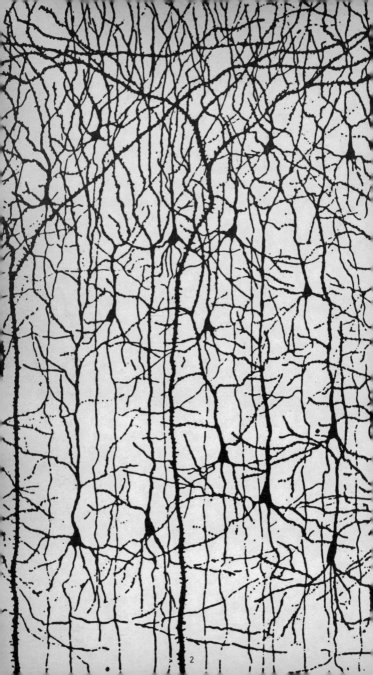

> We shall not cease from exploration
> And the end of all our exploring
> Will be to arrive where we started
> And know the place for the first time.
>
> —T. S. Eliot, *Little Gidding*

PREFACE

To know one's self is in essence to know one's brain—to probe such issues as consciousness, creativity, intelligence. It has been said that the brain, understanding itself, remains mankind's supreme challenge. This book is a progress report of sorts. It is hardly comprehensive; in particular, it does not cover many aspects of psychology. The title of this book can be taken literally: we hope to acquaint the reader with what is inside the brain and convey a sense of how it works. Mapping the cerebral cortex has yielded a great deal of information about how some parts of the brain have developed specialized areas during evolution: not merely the maps of the sensory worlds, but the locations of those parts of the brain which give us our language abilities, our ability to direct our attention, our ability for music, and our ability to read this.

Yet locating the head office for such abilities does little to reveal how the individual nerve cells bring off these feats. Exploring the neuron (a nerve cell by another name) shows how information is processed by electrical and chemical means, how circuits of neurons are wired together to perform more elaborate tasks, such as telling vertical from horizontal. Each neuron by itself turns out to be an elaborate computerlike machine. With many of them working together, complicated patterns can be analyzed and decisions made.

Despite one of its major themes, this book is not really about neurosurgery. As generations of physiologists have known, one of the best ways to illustrate normal function is to show how things go wrong. Epilepsy, strokes, and tumors have all contributed much to our knowledge of "what's where" in the brain. During operations on epileptics to control seizures, electrical stimulation of very local areas of brain can be performed. This "mapping" has provided science with some of the most detailed knowledge about how the different regions of the brain specialize. In this book, we follow an epileptic, Neil, through such a day of neurosurgery and science. We discuss other diseases of the brain in the process of illustrating how neurons work. Thus we are not setting out to survey the scientific knowledge of schizophrenia or Parkinson's disease; rather, we are seeking to illustrate how normal aspects of neurons can go wrong. That we manage to cover quite a number of diseases in this manner reflects the extent to which the scientific quest for fundamental knowledge has proved broadly useful in illuminating disease mechanisms.

Our nonscientist students have convinced us that many aspects of brain research can be digested by the general public. We have been teaching a noncredit course about the brain and have found that there are many areas of neuroscience which can be presented in considerable detail without running into technical barriers requiring preparation in physics or chemistry. We would like to thank our many "students" who contributed suggestions and encouragement. Through them we have gained an appreciation for the varied motivations, backgrounds, and expectations of general readers.

This book, however, is not a textbook for such a course. A textbook writer sets out to convey a certain body of information in a well-organized and accurate manner, usually without the emphasis and comment characteristic of science journalists. One simply assumes that the reader will be impressed by the facts, once they are properly digested; however, one can also assume the presence of a teacher to maintain continuity and that students will reread and study the material. A general reader who tries to get through such a textbook has to be quite disciplined to get beyond the second chapter. As science journalists know, it is not merely a problem of translating technical material into more accessible language: there will always remain some hurdles, different for different readers, and one must motivate the unaided reader over them. Our device to maintain continuity is "Neil": thanks to our familiarity with the neurosurgical operation for epilepsy, we were able to create a central event and a main character, which we periodically revisit. We have also intentionally postponed (at the expense of a more logical organization) some of the more technical material until after the reader has had a chance to become familiar with the general flow of the story. We hope that when we are telling readers more than they really wanted to know about a certain subject, they will understand that they can simply skip ahead to the more familiar material at the beginning of the next chapter. We have tried to make most chapters capable of standing alone.

Every field of endeavor has its own specialized vocabulary, and neuroscience is no exception. First-year medical and graduate students spend months learning all of the names. This can be one of the pleasures: one book reviewer said of a science journalist that he "is captivated by the terminology of high energy physics and obviously enjoys the mystery and incomprehensibility of it all."[1] Our effort to avoid terminology for its own sake, while resulting in a more compact book, does lead to some curious omissions: we have discovered that our manuscript does not mention the corpus callosum, as we approached lateralization of function to left and right sides using evidence other than the popularized split-brain studies.

It is sometimes difficult to get behind the facade of facts, but science has its other pleasures. As one science journalist noted, "I value physics, not for how well it explains the world, but much as the irreligious loved the Latin mass, for how it sounds." Although they seldom talk of such things in print, neuroscientists are excited about the picture of the brain which emerges from their collective research. Sometimes they are impressed in the manner of the Latin mass: it is important how it sounds, how "elegant" it is (contemplating science can be like listening to Bach). Sometimes they are excited by the potential applicability of the knowledge. And

---

[1] Superscripts refer to the *Notes* at the end of the book where we provide literature citations and/or elaborations of the text in more technical language.

often, neuroscientists are intrigued because their findings help them to understand their own personal experiences.

What this book seeks to accomplish is to convey the sense of adventure felt by those engaged in exploring the brain, to show how human intelligence arises out of the varied specializations of the brain, and to demonstrate that these specialized regions are composed of millions of individual neurons whose electrical and chemical properties can be analyzed and understood by neuroscientists. Our present understanding is quite imperfect, and we have also tried to show some of the dilemmas which this poses in the medical and surgical care of those with brain disorders. Unfortunately, not all patients with neurological illnesses recover as well as those we have chosen to illustrate.

Most of the people in this book, whether patients or doctors, are really composites of a number of people whom we have known. Where a case history does represent a single person, the name has been changed. An exception to this is Fred, the patient who awoke one morning to discover that he had lost his ability to read, although his vision was otherwise perfect. He is Fred H. Calvin, the father of one of the authors, and he provides a good example of how patients can recover functions lost following a stroke: he is one of several dozen nonscientists whom we must thank for reading earlier versions of our manuscript. Susan M. Johnston edited numerous drafts and coordinated us all; her forthright judgment and sense of style have been greatly appreciated. Phyllis Wood cheerfully illustrated the book for us.

At this juncture, it is customary to thank spouses and children for their forebearance during the ordeal of book writing. That they have all survived our preoccupation without noticeable harm may reflect the fact that much of the writing was done in airplanes and airports. As our wives are also neuroscientists on the University of Washington faculty, we have gotten much help from them. Dr. Linda Moretti-Ojemann contributed some choice case histories from her neurology experience. Dr. Katherine Graubard alternated neurophysiological criticism with enthusiastic encouragement. Together, they provided one piece of advice that was especially valuable. As we all sat around after a memorable Thanksgiving dinner, they spent hours persuading us to abandon our previous textbook-style draft of this book (which was two years in the making) and to start all over again, using lots of dialogue and case histories instead. Taking their advice has immeasurably improved the book and spared readers many hardships. We thank them.

Seattle

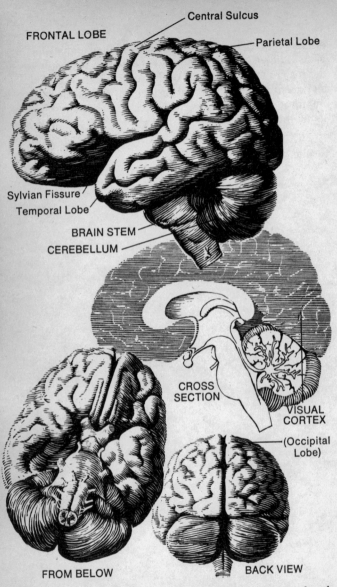

FRONTAL LOBE

Central Sulcus

Parietal Lobe

Sylvian Fissure

Temporal Lobe

BRAIN STEM

CEREBELLUM

CROSS SECTION

VISUAL CORTEX

(Occipital Lobe)

FROM BELOW

BACK VIEW

Different views of the human brain, by Anthony Ravielli (reproduced with permission from *The Human Brain, Its Capacities and Functions*, by I. Asimov [New York: Mentor Books, 1963]).

1

Watching the Brain at Work: Mapping the Cortex

The neurosurgery operating room is a little unusual even from the outside. On its swinging doors today hangs a prominent sign proclaiming, "Quiet Please, Patient Awake." Most surgical patients get to sleep while the OR crew works, hence the reminder to save the OR jokes for another day.

The OR door is a heavy, solid door that does not push open easily except by the surgeon's trick of backing into it. Surgeons have to do that anyway, to keep from touching it with their dripping hands, fresh from the scrub sink outside. Nondripping types sometimes just heave with their shoulders.

Inside, there is all the usual paraphernalia. Tables covered with unironed green tablecloths, the sterile drapes on which the usual array of instruments have been laid out by the scrub nurse. There is the anesthesiologist's cart with air tanks and respirator devices, even though general anesthesia is not being used today. And today there is even more than the usual quota of TV and electronics gear.

The patient is lying on his right side, a doughnut-shaped pillow under his head so that his right ear won't get sore. Neil will be lying in this position all day, although there are always breaks in the operation when he will be able to move around a little on the table and get comfortable again. The operating table is a well-padded couch with infinite variations in possible positions.

The surgeon is looking down on the left side of Neil's head. Earlier in the morning, a U-shaped incision was made in the skin and the scalp folded back over Neil's left ear. That's all the local anesthetic was needed for, since the deeper tissues are not particularly sensitive to pain. Getting through the bone is noisy but not painful. A piece of skull about the size of the patient's hand was removed, using air-driven drills and saws. The scrub nurse has the piece of skull lying on the instrument table, carefully wrapped up in moist sterile gauze, as it will be replaced in Neil's head at the end of the operation, along about dinnertime.

By this time in the middle of the morning, the neurosurgeon has also made a U-shaped incision in the dura, a tough membrane which covers the brain like a thin skin. Like the scalp flap, it too has been folded back and kept moist. For the first time, the brain itself has been exposed to the light of day, or, more precisely, to the big OR lights.

"Neil, now's a good time to move around a little. Is that comfortable for you? How are you feeling?" asks the neurosurgeon.

"That's a bit more comfortable. Oh, I'm feeling okay. How far along are you?" Neil answers.

The OR during an epilepsy operation.

"We're through with all the big mechanical parts of the surgery, Neil. Now we'll begin checking out the brain. Remember to let me know if you feel anything."

The neurosurgeon presses gently on the soft brain in various places, exploring to see if the brain tissue feels unusually tough, which might indicate a tumor, scar, or other pathological change in the brain tissue. Neil does not feel anything, which is perfectly natural, as the brain itself is insensitive to pain or touch; it is not equipped with the skin's type of transducer[1] nerve cells, which specialize in sensing touch.

Fifteen years ago, Neil suffered a skull fracture in a car accident. He now has a type of epilepsy which the usual drugs have been incapable of controlling. The reason for the operation is that a long workup has indicated that the epileptic seizures seem to start in a particular area of the brain, the left temporal lobe, located just in front of Neil's left ear.

Neil is quite interested in all of this, and indeed is having an experience which few people will ever have. Neil is about to learn the precise areas in his own brain which control the movement of his hands and face, that receive the sensations from the transducer neurons in his skin; those areas of his brain which he uses to lift a fork, dial a telephone, play the piano, or speak to the neurosurgeon.

Neil is an engineer, a graduate of MIT. More than most patients who go through this operation, he can conceive of the human brain as something akin to a very elaborate computer, operating on electrical principles, with different regions of the brain specializing in different functions. Although Neil's brain has the same general plan as the brain of any other human being, the details of his brain differ from all others, just as his face is unlikely to be identical to anyone else's. The neurosurgeon needs to know what's where in Neil's brain, and he can't tell by merely looking at the surface of the brain.

"Neil, we are now getting to the part of the operation I told you about earlier, where I am going to stimulate the brain electrically. You're probably going to feel something. Tell me where you feel it."

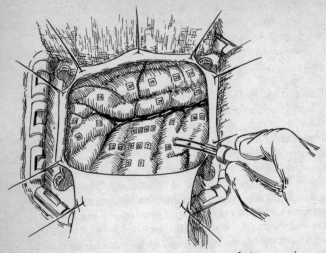

Neurosurgeon's view of Neil's brain during stimulation mapping.

The neurosurgeon picks up a penlike device with two silver wires on its tip. Connected to an electrical stimulator, the device is controlled by an electrical engineer looking down from behind the windows of the gallery above the OR. When the two silver wires are touched gently to the surface of Neil's brain, several milliamperes of current flow through the brain, and then the wires are lifted off.

"Hey, I felt something on my face," Neil says with some surprise.

"Did your face move first?"

"No, it just felt funny, like tingling."

"Well, let me try another spot." The neurosurgeon momentarily touches the wires to an adjacent part of the brain, slightly in front of the first spot.

"My mouth moved!"

The anesthesiologist, who has been watching Neil's face carefully, says, "It was just the right corner of his mouth. It pulled up for a second, and then relaxed."

The neurosurgeon places a little piece of sterile paper, with a number printed on it, on the brain at the site where the *mouth movement was evoked to mark the spot temporarily. The electrical stimulation is then applied to another site, farther away from the ear and closer to the top of the head.*

"Hey, my hand moved! Just like someone else moved it for me!"

Which is, of course, exactly what has happened. The part of Neil's left brain touched by the stimulating wires is part of the brain essential for movement of his right hand. Neil uses it whenever he wants to move his right hand. The electricity applied by the neurosurgeon simply bypasses the voluntary control and directly initiates the hand movement.

That the left side of the brain controls the right side of the body is something that was known even in the ancient civiliza-

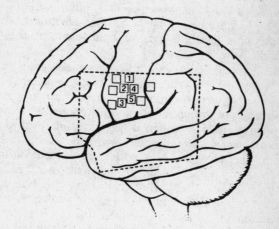

Identification of Neil's left motor and sensory cortex.

Numbered sites: motor or sensory responses to stimulation
1. Hand movement on right
2. Cheek movement on right
3. Tongue movement on right
4. Cheek sensation on right
5. Tongue sensation on right

Open sites: No motor or sensory responses to stimulation
Dotted area: Limits of exposure of Neil's brain (as on page 4).

tions. The great Greek physician Hippocrates observed that after injuries to one side of the head, it was often the opposite side of the body which became paralyzed or might be involved in a seizure. That there is a very orderly map in the left brain for movements of the right side was discovered only in the nineteenth century. Its details were described by a British neurologist, Hughlings Jackson. As the neurosurgeon moves the stimulating wires toward the top of the head, the sequence of face to hand will progress to chest and hips. More detailed "mapping" using this technique will reveal thumb and forefinger areas within the hand area. More space in the brain seems to be devoted to the fingers than to the leg.[2]

Parallel to this "motor strip" controlling muscles is a strip where sensations are evoked. This "sensory strip" is just to the rear of the motor strip. Marching up it with the stimulating wires would have caused Neil's tingling sensation to move with the same general progression from face to hand to the rest of the body.

The motor and sensory strips have an average arrangement over many patients, which allows maps to be made for textbooks. But there is considerable variation between people in the details of this organization, much more than is apparent just from visual inspection of the brain's surface infoldings. A deep infolded groove in the brain, the central sulcus, separates the motor and sensory strips in the usual textbook pictures of an "average" brain. Neil doesn't have that anatomical dividing line. After Neil's electrical-stimulation mapping has been completed, the numbered tags indicating the motor strip lie on the surface of the brain right next to sensory points; no infolded region separates them. It is not presently known whether such detailed differences are important, perhaps reflecting the differences between the clumsy and the highly coordinated. But the differences in brain surface anatomy are certainly there, and that is one reason why neurosurgeons try to map out the brain physiologically in an awake patient, who can help by reporting what happens.[3]

Below the motor and sensory areas, across a groove (the sylvian fissure) that marks the top of the temporal lobe, is a small region where the electrical stimulation will cause Neil

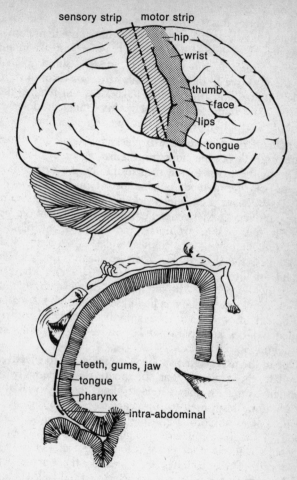

Right brain's sensory and motor strips. Dotted line through sensory strip shows plane of section revealing details (below) of the average patient's sensory map (as modified from Penfield and Jasper 1954). The motor-strip map is similar in sequence (one difference is the neck, which is between face and thumb on the motor map) but somewhat different in the relative magnifications. For example, the hand (excluding the fingers) has three times as much motor representation as sensory representation.

to report buzzing noises: the auditory receiving area. If the left visual cortex located in the back of Neil's brain were to be simulated, he would report flashes of light. Those light flashes would appear in the right side of Neil's vision, for like motor function, vision is wired up in a crossed fashion so that everything we see to the right of our center of vision goes to the left brain. Since the surgery today does not require the back of Neil's brain to be exposed, this cannot be done in Neil but it is well known from studying other patients.[4]

There are many methods which researchers use in animals to map out the connections from sense organs to the brain, and from the brain to the muscles. They are much more precise than the methods suitable for use on humans in the OR. As we have seen, electrical stimulation of the brain's surface produces rather general effects: the tingling of the face is often diffuse, the involuntary movement of the hand is often uncoordinated, the sounds evoked are noises rather than tones, the light flashes have no form. They are very helpful for the neurosurgeon in evaluating what's where in this particular person, but such crude electrical stimulation activates millions of nerve cells indiscriminately. Laboratory equipment allows one to see a limited, selective, orderly sequence of cell activity in monkey brains when the monkey is reaching for a banana.

One of the reasons that applied electricity can so effectively stimulate the brain is that the brain runs internally on electricity, much as does a modern digital computer. Rather than burning fossil fuels to run generators, humans generate electricity by transporting glucose (one form of sugar) in the blood to the brain, where the glucose runs an elaborate metabolic pathway. One effect is to charge batteries in each nerve cell. These batteries then provide the energy with which electrical computations are performed. So-called "brain waves" are an indirect indication of the brain's electrical cycling processes and they play a major role in analyzing Neil's epilepsy.

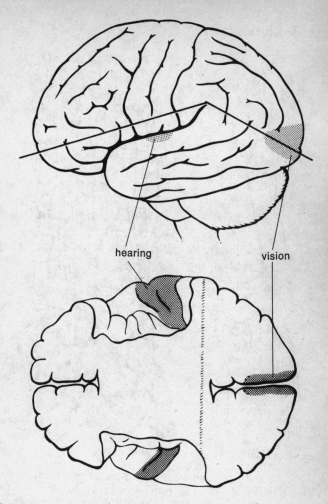

Cortical maps for hearing and vision. The bottom figure is looking down on the brain, as cut along the lines indicated in the top figure. Note that hearing lies buried in the sylvian fissure, on the top of the temporal lobe. Vision is on the inner side of the cerebral hemisphere, although some visual-association areas (not labeled) can be seen from the side view.

LEG AREA IN MOTOR STRIP

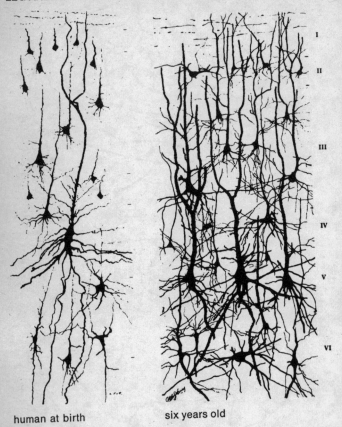

human at birth   six years old

The leg area of motor cortex has some of the largest neurons in the brain. This figure, and similar ones opening later chapters, are from the monumental work of J. N. Conel; they contrast the brain at birth with how it has changed by six years of age.

2

Living Better Electrically: Exploring the Neuron

About this time in midmorning, the electronics technicians get very busy and several more doctors appear. Rather than coming scrubbed into the OR, the doctors wander into the glassed-in gallery upstairs, which looks down into the OR. They are neurologists who specialize in the interpretation of the electroencephalogram, more commonly called EEG or "brain waves." They first analyzed Neil's brain waves months ago by recording minute electrical voltages from sixteen places on the skin of Neil's head.

Epileptic seizures are often preceded by abnormal EEG voltage patterns; indeed, there are often abnormal EEG patterns in epileptics even when a seizure is not imminent. For the muscles to start jerking, the abnormal electrical activity must involve wide areas of the brain. In many epileptics, seizures originate from only a small region and then disrupt the surrounding brain, which is otherwise trying to go about its normal business. Locating this "epileptic focus" is the big problem.

In Neil's case, the scalp EEG suggested that this epileptic focus was in the left temporal lobe, which is in front of the left ear and below the motor and sensory strips. Since that portion of the brain can often be removed with little obvious effect upon a patient, Neil is willing to trade a potential deficit for the chance that his seizures can be brought under control. Had his epileptic focus been in the middle of the motor strip, it would have posed a dilemma: removing the left motor strip would have produced permanent paralysis of the right side of the body. Better seizures than paralysis.

The neurosurgeon gently places the tips of eight silver wires on the surface of the brain and hooks the wires up to a cable leading to the EEG machine in the gallery. A TV camera above the EEG machine is connected to the TV screen downstairs in the OR, so that the neurosurgeon can see the results too. These recordings from the surface of the brain are much more accurate than the EEG obtained earlier from the skin, where the voltages had to travel through the dura, the bone, some muscle, and then the scalp.

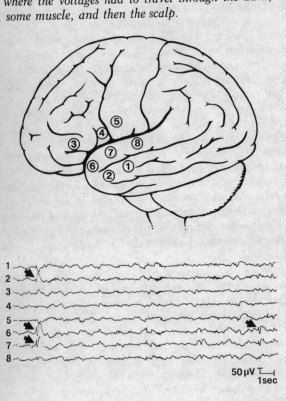

50 µV
1 sec

Neil's EEG, as recorded from eight points on the surface of the exposed brain. This is not a seizure; the sharp deflections are, however, the trademark of a resting epileptic focus. By contrasting the recordings from different sites on the brain, it can be deduced that the epileptic focus is in the tip of the left temporal lobe.

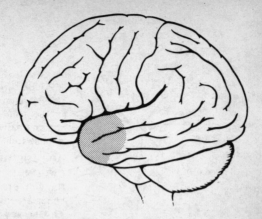

Neil's epileptic focus at the tip of the left temporal lobe.

The EEG voltages are a product of the electrical signals from millions of nerve cells, mostly from those near the surface of the brain. The EEG is an indirect indicator of their overall levels of activity, about the way that one could judge the overall daily activity patterns of the inhabitants of a city by listening to traffic noises. What, however is the individual nerve cell doing with electricity? Nerve cells have two major tasks: computing, and then speeding the results of the computation to the far end of the nerve cell so that they can be passed on to another nerve cell. Both tasks run on electricity.

The individual nerve cells (also called neurons) are shaped like leafless trees, with branches and roots separated by a long trunk. Sometimes the trunk, called the axon in a nerve cell,[1] is only one millimeter long. Some neurons have an axon so long that it can reach from the tip of the toe to the base of the brain. An electrical signal speeding along this axon tells the brain that something has touched the big toe. This electrical signal is called the impulse.[2] It is only 1/10 of a volt (more than a thousand times smaller than household electricity), it lasts only 1/1000 of a second (quicker than most camera shutters), and it races along the axon at speeds as high as 500 kph (300 mph).

THE NEURON

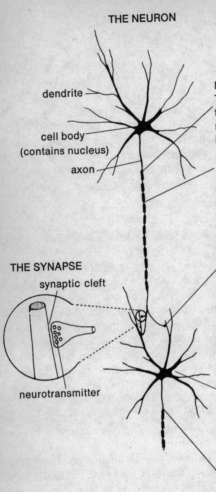

dendrite

cell body
(contains nucleus)

axon

THE SYNAPSE

synaptic cleft

neurotransmitter

IMPULSE INITIA-
TION at axon's ini-
tial segment sets
rate and pattern of
impulse train.

IMPULSE REPLI-
CATION at node of
Ranvier provides
"booster station"
and speeds im-
pulse along to
axon terminal.

SYNAPTIC
TRANSMISSION
produces voltage
change in down-
stream neuron
when impulse re-
leases neurotrans-
mitter packet from
axon terminals.

DOWNSTREAM
NEURON inte-
grates positive and
negative voltages
from thousands of
upstream neurons.

IMPULSE INITIA-
TION from down-
stream neuron en-
codes net voltage
result as another
impulse train. . . .

The components of a neuron, with summaries of their functions.

More often, the race is a relay race: there is a chain of nerve cells connecting the transducer in the skin with the brain. The impulse is passed from one nerve cell to another by an electrically triggered squirt of a chemical. In response to the momentary change in voltage, a chemical is secreted from the end of one nerve cell and diffuses to the next nerve cell in the chain. This junction between two nerve cells is called a synapse. The chemical, called a neurotransmitter substance, in turn produces a change in the voltage of the downstream nerve cell.[3] Many types of drugs, such as painkillers and tranquilizers, interfere with this chemical process connecting two nerve cells with each other at the synapse, and thus enhance or reduce the strength of the connection.

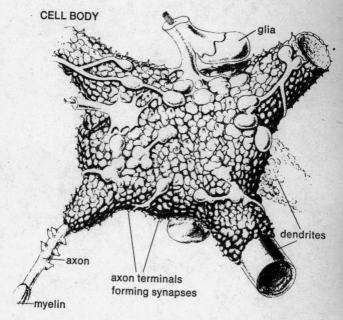

CELL BODY

glia

axon

axon terminals forming synapses

dendrites

myelin

At the junction of the dendrites and the axon, the cell body is encrusted with synapses from upstream neurons. Less than 10 percent of the synapses are seen here, as most are on the dendrites. Glia are not neurons; they are cells which support and nourish neurons (redrawn from Poritsky 1969).

But this chemically evoked voltage change in the downstream cell does not usually cause an impulse by itself. The second neuron in the chain from transducer to brain is typically located in the spinal cord. There are hundreds of transducers in the skin and muscle converging upon this cell in the spinal cord, plus thousands of inputs from other neurons in the brain and spinal cord. Some produce negative voltage changes ("inhibition"), others produce positive changes ("excitation").

As in a checkbook, the balance is what counts. If the voltage balance is big enough in the positive direction, another impulse will be triggered and will speed along the axon of the second nerve cell, often heading up toward the brain. The balance also determines the rate at which impulses can be produced, which varies between 0 and 1,300 impulses

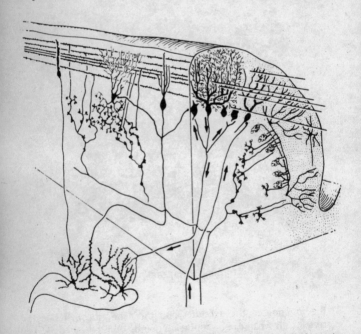

This example of one of the simpler circuits of neurons in the brain is from the cerebellar cortex. The cerebral cortex seems considerably more complicated (modified from the work of C. A. Fox 1962).

each second. Each nerve cell is thus a simple computer, adding and subtracting influences from many inputs, sending its new message on to many other cells. Chemicals are very important in producing the electricity, just as chemical reactions in batteries may be said to run a pocket calculator or a wrist watch. Chemicals also may slowly change the strength of the connection between two nerve cells. But it is with electricity that nerve cells add and subtract, and by using electricity that they speed their messages along to the next nerve cell. The EEG and "evoked potentials" are noisy by-products of all this electrical processing of information, but clinically useful ones.[4]

The EEG is not, however, an average of the electrical activities of the brain beneath the electrode. The closer that the neurons are to the electrode, the bigger their contribution. The cerebral cortex is only a few millimeters thick (less than two typewriter spaces). It may be divided into layers (usually six, numbered I-VI but sometimes subdivided even further, e.g., layer IVc). The largest neurons of the cortex are often found in the deeper layers, such as layer V. The inputs to the cortex often prefer certain layers, e.g., the messages from the eyes arrive only in layer IVc of visual cortex. The most poorly understood layers are the ones closest to the brain's surface (I, II, and III), and they are the ones which contribute most heavily to the EEG. Yet despite the lack of detailed understanding about the origins of the EEG in terms of the activities of neurons, a practiced eye can recognize normal and abnormal patterns in the electrical fluctuations of the EEG. Such empirical knowledge is essential to the analysis of where Neil's epilepsy originates.

LANGUAGE CORTEX

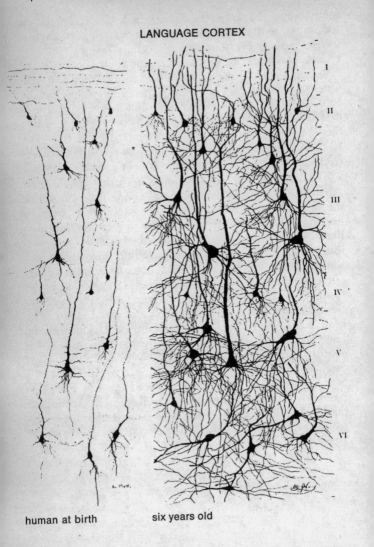

human at birth   six years old

3

Reading, Writing, and Speaking: Where Does Language Live?

The rapid voltage fluctuations characteristic of a "resting" epileptic focus are seen all along Neil's left temporal lobe, much as the earlier scalp EEG had suggested. However, they are seen far enough in the rear of the temporal lobe to cause the neurosurgeon some concern, as this area and the adjoining parietal lobe are very important for language. Just as one does not remove the cells which control movement, one does not remove language portions of the brain to control seizures: the cure would be worse than the disease. But what are the boundaries of the language part of Neil's brain? Do they overlap the epileptic focus, or is there a strip in between of healthy but nonlanguage brain tissue?

"Well, Neil, so far, so good. Now we'll set up the slide show for you. Everything okay?" asks the neurosurgeon.

"Oh, I'm fine. And if it's the same slide show as before, it's about the easiest test that I've ever taken."

Neil, who had been resting quietly, listening to everyone discuss his EEG interpretation, is now put to work again. The night before, Neil was trained on a simple test of language and memory. The apparatus for it is now set up in the OR. There is a slide projector and a small projection screen. The slides are shown sideways on the screen so that Neil, lying on his side, sees them normally. Everyone else in the OR can be seen tilting their heads sideways to watch.

The first slide is a sketch of a star. Neil says, "This is a star."

The next slide is just the number 84. Neil knows that his task is to count backward by threes, so he begins "84, 81, 78, 75 . . ."

The third slide just has the word "Recall" on it. Neil stops counting and replies "Star."

Patient viewing slide show (on rear-projection screen) while language cortex is being electrically stimulated. Patient's responses are tape-recorded.

The fourth slide is a sketch of a house, so Neil says, "This is a house." Another number comes up, and he again counts backward by threes to distract him from rehearsing the name that he will need to recall six seconds later. Over and over the three-slide sequence repeats, using sketches of many different common objects.

This series of tasks is hardly difficult, particularly with the previous night's practice, and Neil would ordinarily make few mistakes. But during the surgery, electrical stimulation is applied to the brain during certain parts of the sequence, not applied during others. The electrical stimulation is weak enough that Neil cannot tell when it occurs. Unlike the stimulation of motor strip, stimulation of speech and language areas does not evoke speech; rather, it confuses things so that a small area of cortex does not function properly.

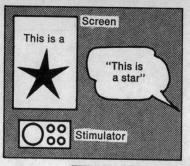

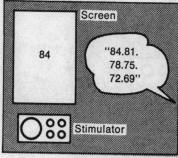

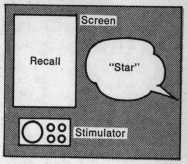

A typical three-slide sequence used during language testing in the operating room. Without stimulation, the patient has no trouble naming the drawing and can recall the name following six seconds of distraction (counting backward by threes, cued by the second slide).

When the language cortex is stimulated, the patient's ability to name the object is impaired even though he can still talk.

"This is . . . it's a . . . you know, it's a . . . I just can't . . ." Neil can talk, but he cannot find the right name for the object. Interestingly enough, after the electrical stimulation is turned off and Neil has counted backward by threes and the "recall" slide has appeared, he may now be able to recall the correct name.

This test of naming helps to identify the language areas of the brain for the neurosurgeon. The neurosurgeon moves the silver stimulating wires around over the exposed surface of the brain during the test, keeping track of the spots where the stimulation disrupts naming by laying another little numbered paper tag on the spot.

Sometimes, the electrical stimulation is given while Neil is counting backward by threes, just after Neil has correctly named the sketch of the object. Neil continues counting backward just as if the stimulation had no effect. However, when he tries to recall the object's name a few seconds later after the stimulation is turned off, he now has trouble remembering. Only stimulation at some points surrounding the language area has this dramatic effect of wiping out the memory of the sketch.

The language area just defined, by electrical stimulation during object naming, is one of the two language areas known from the studies of stroke patients. More than a century ago, the German neurologist Karl Wernicke described patients who talked nonsense but with good grammar, mak-

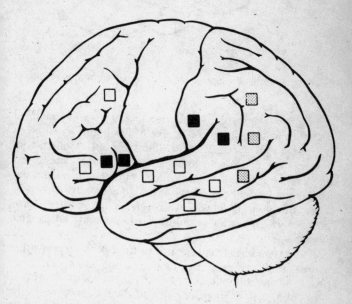

Sites in Neil's brain where stimulation produced naming errors (filled squares) and where stimulation during the distraction produced memory errors (dotted squares). The open squares are sites where no effects were produced.

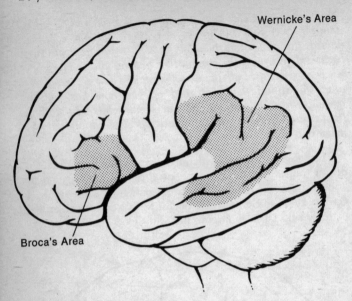

Traditional cortical language areas, inferred from the results of strokes.

ing up new words so as to talk in a jargon. A woman named Blanche, with damage to this area of her brain, when asked her name said, "Yes, it's not Mount Everest, Mont Blanc, blancmange, or almonds to put in water . . . you know, you be clever and tell me." Such patients often fail to understand what others say to them, as well as talking nonsense themselves. They are inclined to talk excessively. These patients have strokes,[1] which destroy portions of left temporal and parietal lobes.[2]

There is another language area in front of the motor strip, named after a French neurologist of the same era, Paul Broca. Destruction there causes a patient to speak with a telegraphic style of language, with all the little words missing and with little grammar. Talking about a movie, such a patient said, "Ah, policeman . . . ah . . . I know! . . . cashier! . . . money . . . ah! cigarettes . . . I know . . . this . . . beer . . . mustache . . ." This patient understands what others say, knows he cannot talk correctly, and is frequently very frustrated by it. With more extensive damage, speech may be reduced to only a word or two, used over and over again for everything. Such was the situation in the patient under Broca's care in whom he first recognized this type of aphasia. Broca was on the staff of a large nursing home where this patient had lived for some years following a stroke. The patient's name was LeBorgne, but he was known to all as Tan-tan, for that is all he could say: "Tan-tan . . ." At autopsy after Tan-tan died, Broca noted the damage to the left frontal lobe and related this to the language defect.[3]

This frontal language area is usually considered to be concerned with the output of language; the larger area of the parietal and temporal lobes is thought to be more concerned with the understanding of language. Thus the speech changes with the damage to the frontal area are often called "motor" or "expressive" aphasia, while those after damage to parietal and temporal lobes are called "sensory" or "receptive" aphasia. ("Aphasia" literally means "no speech," but in practice the term is applied to someone with any kind of language defect after brain damage.)[4] Extensive damage to both frontal and parietal-temporal language areas, as in a large left-hemisphere stroke, produces obvious deficits in both speech output and understanding, a combination called "global aphasia."

More subtle language disturbances have been described after small strokes damaging only part of one language area, or the connections between areas. Some patients with persisting damage to a small part of the parietal-temporal language area show difficulty only in coming up with the names of things, a condition known as "anomic aphasia." Such a patient, when shown a key, may say, "That's a, uh, I know that, it's, ah, a thing you put in a lock." Otherwise his speech is normal. Although anomic aphasics have difficulty only in naming, most other aphasic patients have naming difficulties as well as other speech deficits. Because such problems with naming simple objects are common to all types of aphasia, neurologists and neurosurgeons find this a useful test to be sure that language is intact and to learn where the language area is located during operations such as Neil's.

In another type of aphasia, spontaneous speech is usually present but there is a severe difficulty in repeating back spoken words. This deficit occurs with damage to a portion of the parietal-temporal language area, and is called "conduction aphasia."[5] The reverse of this, intact repetition but difficulty with initiating spontaneous speech, is occasionally seen after damage to the connections between the major language areas ("transcortical aphasia").

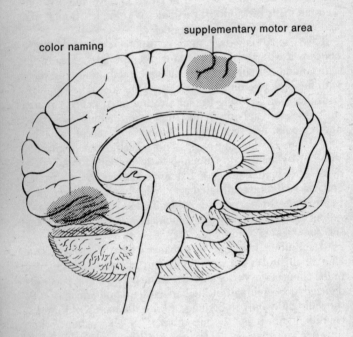

Sites on the inner surface of the left hemisphere where strokes will produce an inability to name colors. Also shown is the location of the supplementary motor area, one of the few places where brain damage may leave a patient temporarily mute.

An even more unusual aphasia is an impaired ability to name colors. In these patients, the damage is on the inner side of the hemisphere, just below the area of cortex involved in vision.[6] Presumably this damage disconnects visual cortex from language cortex, which particularly affects color naming because colors are one of the few pure associations between naming and a visual input. There are few color clues from shape or usefulness; few other words help in identifying colors.

Rarely is a patient completely mute. Stroke patients can usually produce a word or two, even with severe damage to language areas. Such patients may occasionally surprise their families by uttering a long string of invective: emotional speech is generally better preserved in severe motor aphasia, as is laughing and crying.

We do not yet know the areas of the human brain which are important for emotional speech. In monkeys, the specific vocalizations produced by each species to attract mates or defend territory seem to be analogous to human emotional speech; in the monkey brain, these vocalizations depend upon the inner face of the hemisphere just in front of the foot area on the motor strip, a region called the supplementary motor area.[7] In man, damage to that area does temporarily render the patient mute; fortunately, essentially full recovery occurs in weeks. This area would be the logical candidate for human emotional speech. Emotional speech seems to be less lateralized to left brain than other language; indeed, some features of it, such as laughing, have been identified more often with right-brain function.

Despite their severe deficits in language, many aphasic patients are otherwise normal. Motor and sensory functions, hearing, and vision can be intact, and the patient may know that he does not speak correctly. When the problems of getting information into and out of the patient (by some mechanism other than speech) can be solved, tests often show normal ability to reason and make inferences. Brain damage leading to aphasia does not necessarily impair the nonlanguage aspects of complex thought. An aphasic artist may still draw, an aphasic musician may still compose. We do not know just how severely the language aspects of complex thought are

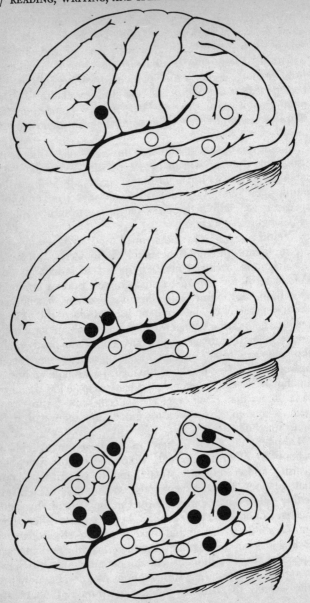

impaired by this damage, largely because of the problem in testing such patients with impaired language. It has been argued that, with maturity, thoughts expressed in words are increasingly internalized into "inner speech." Young children tend to think out loud; adults keep those words to themselves. Such internalized phenomena are not easily studied, but it has been suggested that some types of aphasia severely impair this inner speech, especially aphasia after damage to Wernicke's area.

In the preceding section we have presented the general pattern of language organization, with a frontal-lobe area for language output and a temporal-parietal area for language understanding.

This pattern may be quite different in young children or left-handed persons, as we shall discuss in a subsequent chapter. But even in the average right-handed adult, there seems to be considerable individual variation in the exact cortical location of language. For example, the sites where electrical stimulation of cortex alters naming are seldom exactly the same from one patient to the next. Only the rear one-third of the frontal language area, apparently necessary for the motor output of speech, seems to be relatively uniform between patients. In any other spot in the usual language areas, some patients will show naming changes indicating that they have language there, others will not. The individual pattern of language localization seems to be quite variable, perhaps as variable as our individual appearance. There are suggestions that overall verbal ability may correlate roughly with some patterns of language localization.[8]

Surprisingly, two different languages spoken by the same patient may not use exactly the same areas of brain. Strokes damaging language areas in bilingual patients most commonly impair both languages equally or in proportion to how well they were spoken before the damage. The patient's native tongue, or the language in which he was most fluent, or the

Mapping of language areas in three different patients. Filled circles are sites where stimulation produced naming errors; open circles are sites with no errors. Note how different each patient is from the others and from Neil (page 23).

one used in his surroundings, is most likely to recover first. But an occasional case is a glaring exception. In these cases, the first and perhaps only language that recovers may be one that was not the native tongue, or the most fluent language, or the language of the present environment; rather, it is some other little-used language.[9] Imagine the family's consternation when, after a stroke, Grandmother speaks only Croatian, which she spoke for a few years as an adolescent. And now no one else in the family speaks Croatian! These cases were especially perplexing to students of language until studies were done using electrical-stimulation mapping during naming in two languages. In bilingual patients, there are some cortical sites where naming is altered in only one language, as well as other sites where stimulation alters both languages. There seem to be brain areas used by only one language or the other, and presumably a stroke may destroy the area for one language (perhaps English) and leave the area for another intact (perhaps the area for Croatian). This does not necessarily mean that there are "first" and "second" language areas predetermined genetically. It may be that a large area of language cortex is needed to first learn any language. With continued use, smaller regions may suffice, freeing up other portions of language brain for new tasks, such as learning another language.

Most aphasic patients have difficulty with reading and writing as well as speaking. Indeed, they tend to make the same kinds of mistakes in both spoken and written language. But occasionally brain damage will be followed by quite intact spoken language, but severe impairment of reading and/or writing. Impairment of only writing is quite rare, and usually associated with damage to wide areas of brain. But impairment of only reading is not uncommon. An occasional reading deficit follows left-frontal-lobe damage, above the frontal language area. There the defect seems to be in tracking the printed words, as that area of brain is also involved in voluntary control of eye movements. But more commonly, isolated impairment of reading is due to damage to the back margin of the parietal-temporal (Wernicke) language area. Such damage is thought to disconnect the language area of the brain from visual input.

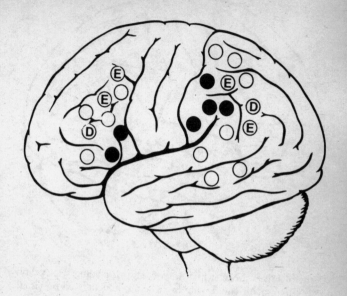

Language cortex in a bilingual patient. Dutch was this patient's native tongue; English was learned as an adult and was the language of his surroundings. Filled circles are sites where naming errors occurred in both languages upon stimulation. Sites marked E showed significant changes in naming errors only in English. Sites marked D showed significant errors only in Dutch. Open circles are sites where no significant errors occurred in either language.

Reading in languages that make use of pictographs, such as Chinese or the Kanji form of Japanese, seems to be less impaired by this type of brain damage (as is the recognition of certain italic-like types of script). Pictographs may be decoded in part by spatial mechanisms in the right brain, discussed in chapter 7. [10]

Aside from this, reading depends on the left side of the brain in most of us. The following patient illustrates this isolated reading disturbance after a stroke damaging the rear portions of parietal and temporal lobes.

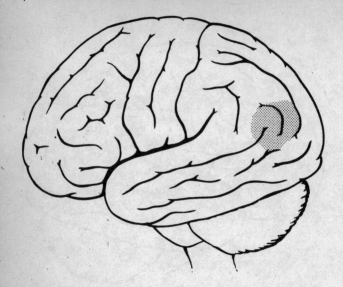

The site of Fred's stroke, which interfered with reading. He had no speech impairment, could write to dictation, and suffered no visual field defects (blind spots).

Fred had a bad headache, quite unlike any other he had ever had before. The next morning, he felt better and got dressed. He went downstairs and put breakfast on, then went out to get the newspaper. Upon sitting down to read it, he discovered to his astonishment that he could not read. He could see the paper, recognize the letters individually, but only the shortest two- and three-letter words made any sense.

Later in the hospital, he was asked to write to dictation. He accurately wrote out in longhand an involved paragraph which was read aloud to him. Asked to read back what he had just written, he couldn't. Only the shortest words made any sense. He would try to piece together longer words but often made errors. His spoken language was normal. He had no blind spots in his visual fields, and his color vision was normal. A computerized tomographic scan[11] showed that his stroke had damaged an area of the left parietal and temporal lobes, *above and behind his left ear. This is to the rear of the language area but in front of the visual part of the brain, an area evidently more important for reading than for spoken language.*

A year later, Fred had recovered most of his ability to read. He tired easily, but could slowly read anything that he wanted (indeed, he read an early draft of this book). Whatever the function lost—reading, muscular control, language—at the time of the stroke, there is typically some recovery of function in the weeks and months which follow. But the extent of this recovery is quite variable. Sometimes, as in Fred's case, there is quite a lot of recovery. Sometimes, especially with more extensive damage, there is only a little recovery, although it may be six to twelve months or more before recovery reaches its limit.

What happens in the brain during those weeks and months is not entirely clear. New neurons don't grow. Wounded neurons, however, return to duty. There may also be a reserve of neurons which can be pressed into action. It is unlikely to be a reserve of unused, spare neurons; a more likely analogy is the army reserve, whose members are called up out of many different occupations in time of need. Skills are widely spread around the population: just because someone specializes in banking does not mean that he or she cannot also function as a truck driver (but perhaps with somewhat less skill than a person who normally specialized in truck driving). Recovery of function thus probably reflects return of activity in damaged but not destroyed neurons and acquisition of secondary functions by other neurons. Indeed, with recovery after damage to language areas, there is now evidence that neurons on the other side of the brain have acquired such a secondary language function.[12]

POSTERIOR PARIETAL

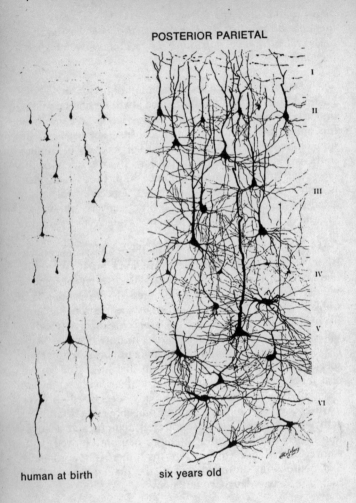

human at birth   six years old

The posterior parietal area is approximately the area damaged in Fred's stroke (previous figure), which interfered with his reading ability.

4

Subdividing Language Cortex: Naming, Sequencing, Syntax, and Short-term Memory

Stroke victims have taught science a great deal: by their own descriptions of their difficulties and from their patient cooperation with neurologists, a rough picture of cortical function has been built up over the years. Epileptic seizures have provided another way of guessing at the function of a particular portion of the cortex, but electrical stimulation of the exposed cortical surface gives more precise localization than strokes or seizures. The exposure of the brain during an epilepsy operation under local anesthesia provides a rare opportunity to explore the organization of the human language cortex, and Neil has agreed to help out.

The "slide show" that was used to define the language portion of the cerebral cortex for the neurosurgeon is now finished. A new tray of slides for the research study is loaded into the projector. A TV camera positioned to watch Neil's face will record the action on videotape for later analysis.

"Neil, we're all set to try those slides where you have to stick out your tongue. All comfortable?" asks the neurosurgeon.

"Sure. I never thought that I'd be on television, sticking out my tongue."

Neil's task, which he rehearsed the night before, is seemingly trivial. Each slide shows drawings of three faces: for example, a face with lips in a kissing position, followed by a face with the tongue sticking out straight, then to the right. Neil is to mimic the three faces in sequence, so he puckers up his lips and then sticks out his tongue straight and then off to the right. The next slide shows further instructions in the form of three more faces. As with the language testing, electrical stimulation is sometimes applied, sometimes not. When the

35

stimulation is applied at certain sites, Neil has trouble producing the sequence of facial movements. A map is thus made of the various sites in Neil's brain where sequential oral-facial movements are disrupted by stimulation.

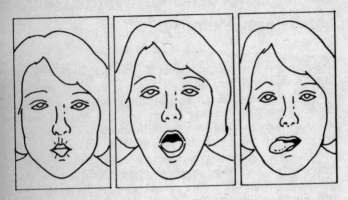

The slide showing the sequence of oral-facial movements which Neil was to mimic.

The movements often involve both left and right sides of the face simultaneously; although only left brain is being stimulated, that is capable of disrupting the sequence of these movements. Stimulation of the right brain, in other patients whose right hemisphere is being operated upon for an epileptic focus, does not do this. Thus, the control of sequential oral-facial movements is also lateralized to the left side, just like language.

Such research was inspired by studies of stroke patients with language difficulties. Many of these stroke victims can mimic a single facial position, such as sticking their tongue out, but they cannot mimic a sequence of them such as Neil can. It was suspected that there was an area in the left brain which is important to the sequencing of motor tasks, probably hand movements as well as face movements. A deaf person who communicates using sign language may be just as impaired by a left brain stroke as a speaking person.[1]

"Okay, Neil. Now we come to the nonsense words. Tell us the letter which changes."

Neil now listens to a series of nonsense words in which one speech sound (phoneme) has been changed: "adma, akma, abma . . ." He identifies the changed phoneme: "k," "b," and so on. This ability to identify phonemes can be disrupted by stimulating certain sites in the language cortex, so that Neil utters an erroneous consonant.

Neil is also asked to read and complete a series of simple sentences which are shown on slides, such as "If my son is late for class again, the principal . . ." When electrical stimulation is applied at certain sites, Neil will make peculiar errors, such as "If my son will getting late today, he'll see the principal." These errors seem to involve only the grammar of the sentence, not its content.

The inability to name objects, but with the ability to speak still functioning ("This is a . . ., is a . . ., I just can't . . ."), is one way to define the language cortex. The study that Neil has just completed is designed to determine whether sites exist which disrupt phoneme discrimination, or grammar, or sequential facial movements, or memory. If so, are these aspects altered from the same or different sites?

A surprising result is that at sites where phoneme recognition can be disrupted, stimulation will also interfere with the mimicry of sequential oral-facial movements. The converse is also true. This finding supports an idea known as the "motor theory of speech perception." Studies have been done of the way manipulation of acoustical cues alters speech understanding; they indicate that understanding of a speech sound has more in common with the movements with which the sound is spoken than with its acoustical properties. Based on those studies,[2] it has been proposed that understanding the spoken word depends upon the brain making an internal model for how to speak the word it hears. The study which Neil has just completed identifies an area of the brain where that internal modeling might occur: this region has a role in both speech understanding and speech production.

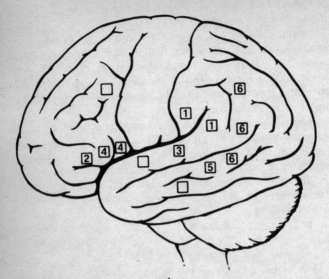

Neil's experimental language study.
1. Sites where sequential movements, phoneme discrimination, and *naming* are altered by stimulation
2. Site where sequential movements, phoneme discrimination, and *reading* are altered (reading was in jargon)
3. Site of grammatical errors in reading (no other errors elicited)
4. Sites with arrest of all language output
5. Site where stimulation caused Neil to read in jargon (no other errors)
6. Sites with memory errors only
   No number: no errors at that site

At other sites in the language cortex, only one of these additional tests is disrupted. Sites disrupting memory rarely show any other effect. Sites showing grammar disruption show no other effects, at least not for any of the aspects of language tested thus far in patients. At the level of resolution provided by surface electrical stimulation, the language cortex seems to put different aspects of language in different places. The exact location of such subdivisions varies from patient to patient (just as the map of the motor strip may vary), but their location relative to each other is generally predictable.

Such findings will likely force a revision of the traditional ideas about brain organization for language which have been inferred from stroke victims; these have emphasized separate regions for receiving language and for producing speech. The new findings suggest that some areas of brain are common to expressive and receptive aspects of language. The major subdivisions of language cortex instead seem to be this common area for sequencing movements and understanding speech, and a surrounding area for memory. Both of these major subdivisions are to be found in frontal as well as parietal-temporal lobes. So it isn't too surprising that careful testing of patients with traditional motor aphasia from frontal damage often shows a subtle defect in understanding as well;[3] that is

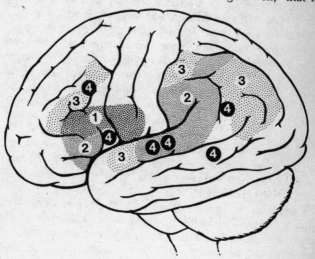

Summary of the experimental language study in four patients, where stimulation:
1. alters all face movement and arrests all speech, a motor area for speech;
2. alters only *sequential* facial movements and also disrupts phoneme discrimination, the area common to speech production and understanding;
3. alters short-term memory for verbal material;
4. alters only the grammatical aspects of reading.

likely from damage to the cortical subdivision common to sequencing face movements and speech understanding. Indeed, damage to this subdivision may underlie all language disturbance. Additional damage, to the memory subdivision, likely leads to a predominantly expressive aphasia. Additional damage to cortex involved in oral-facial motor control probably leads to a predominantly motor aphasia. Language, then, seems to arise in the interaction between these two major subdivisions, with at least some specialized aspects of language such as grammar located at cortical sites between the sequential-movement/speech-understanding and the memory subdivisions.

In the early nineteenth century, neurologists proposed a rather detailed map for functional localization in the cerebral cortex, even down to having a separate area for grammar. Their views fell into disfavor and they were derided as "diagram makers" by those holding a contrary view: to use grammar as an example, the critics would have said that broad areas of brain worked cooperatively to produce grammar, so that no one area was a specialist in grammar. One of the early leaders of the Viennese localization school, Franz Joseph Gall, went on to propose that various regions could be detected by feeling the bumps on the skull, which did not help to keep the idea of detailed maps respectable. While the nineteenth-century maps are not being resurrected, since they were based largely upon speculation rather than observation, the new information from patients such as Neil show that the old ideas were partially right.

Except for the cooperation and participation of patients such as Neil, it is difficult to imagine how such details of the language cortex would be discovered. Epileptic seizures seldom stay confined to such small regions. Strokes may be that small, but matching up symptoms with a brain site is difficult for small strokes: initially, a lot of surrounding brain is made temporarily nonfunctional by pressure (the killed area swells); thus, the initial symptom may be produced by neighboring regions. Later, other areas may take over the functions of the damaged region so that the symptoms disappear, confounding attempts to identify the region's function.

So taking some extra time in the middle of an epilepsy operation is one of the best ways to get the information. Who decides whether to do it? Finally, the patient. But before asking the patient to participate, the researcher must first take a description of the study to a special committee (often called a review committee on human experimentation) whose members are physicians, nurses, and various types of professors and laypersons. This committee asks about the risks (such as prolonging the operation, which might increase the risk of infection). Do the risks outweigh the potential benefits to the patient (knowing more about the functional localization may help the surgeon to better plan the region of the brain to be removed) and to society (for example, potential benefits of new knowledge to dyslexic children or stroke victims)? Can't the study be done in animals instead of humans? (Not language studies.) The investigator must also show the committee the written consent form which the patient will be given to sign. Such committees often ask the investigator to rewrite it, to explain things in simpler language on the written form and to make the risks and potential benefits clearer. Such carefully monitored human experimentation is involved not only in the study of abilities unique to man, but also when new treatments (surgery, drugs) developed in animal research are first applied to patients.

ARM AREA IN SENSORY STRIP

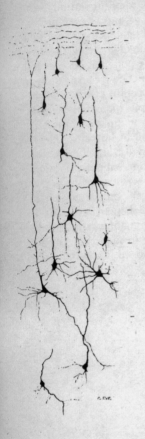

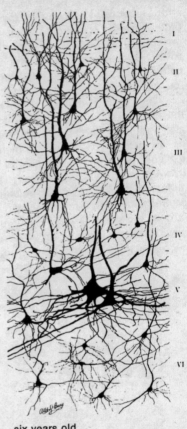

human at birth   six years old

5

Deep Inside the Brain: Compensating for Parkinson's Disease

Down in the OR, the surgical team works in relative silence. The staff in the OR may not be as quiet or as reverent as a concert audience, but it is far quieter than the crowd in the glassed-in gallery, whose voices cannot be heard in the OR unless the intercom button is pressed. The visibility from the brightly lit OR into the darkened gallery is also poor, so the crowd can engage in wisecracks and imitations of the surgeons without serious chance of discovery.

"Is this the first time you've seen a living, working brain?" the neurologist asks one of the medical students standing near the EEG machine.

"Yes, though I could see it better if the surgeon were transparent," complains the student, contemplating the surgeon's back. "I always thought brains were gray and hard. Turns out they are pink and soft. And look at the size of those blood vessels on the surface!"

"Well, the brain does get one-fifth of everything that the heart pumps," comments the second student. "Got to carry it in something."

"That's the middle cerebral artery in the center of the craniotomy opening," notes the neurologist, pointing. "What happens if it gets plugged up?"

The students recognize the standard faculty teaching gambit, the rhetorical question. They respond by reciting to her the whole story of the middle cerebral artery, how it serves the speech and language areas of the brain, and they then rattle off the different kinds of aphasia which can result from blockage of it.

"That's a nice job of covering the surface of the brain, but can you think of any areas in the depths of the brain concerned with speech?" she asks.

43

"Well, the thalamus must have something to do with speech," ventures the first medical student. "They presented a patient with Parkinson's disease in the neurology rounds the other day. He had an operation on his thalamus and was aphasic for several weeks after the surgery."

"Yes, but lots of Parkinson's patients have trouble talking," points out the second student in a dubious tone of voice. "Their voices fade out when they talk on the phone and—"

"Whoa!" interrupts the neurologist. "Suppose you first tell me about the Parkinson's-disease patient. Then we'll discuss treatments for Parkinsonism, and then come back to the aphasia. OK?" The neurologist is off on another teaching tangent.

"This patient said that he couldn't get up out of a soft chair," the first student recalls, "and sometimes couldn't roll over in bed, or cut steak with a knife and fork. He would get stuck in one position and had a hard time getting going again. I guess that he had fallen down several times, too."

"And . . . ?"

"He had a tremor of his hands—no, it was only his left hand—that was most noticeable when he was resting. If you moved his left arm for him, it felt very stiff and would jerk a little—that's 'cogwheel rigidity.' And of course he couldn't move his fingers very fast, and he didn't move his face much, either. His face looked like a mask, just like the textbooks say. His voice kept getting softer and softer as he talked to us. When we asked him to walk around the room, he took little steps and seemed ready to topple over—just as if the top of him was trying to go faster than his legs."

"And what treatments had he had besides the surgery?" queries the neurologist.

"He's currently on levodopa."

And, except for the drug called levodopa and the mention of thalamic surgery, the students' description of the disease is very much the same as that given by James Parkinson in 1817 in his *Essay on the Shaking Palsy*. As so often happens in medicine, it is simpler to identify a disease with a person's name than it is to give it a more meaningful (but longer) label.

For nearly a century and a half, the treatment of Parkinson's disease changed little, although cases became more common. Cases of Parkinson's disease were more frequent following the influenza epidemic after the First World War, particularly in those people who suffered a severe case of encephalitis. By trial and error, some drugs were found which helped some victims, but their side effects prevented many patients from taking them. At the beginning of the space age, many Parkinson's victims were almost as bad off as those at the beginning of the industrial revolution had been.

Many basic science discoveries, however, had occurred in those 150 years which provided the basis for a major advance in the therapy of Parkinson's disease. Anatomically, the brain had progressed from being *terra incognita* to being well labeled. Understanding the function of those hundreds of anatomical subdivisions is a much harder task, and the success story with Parkinson's disease illustrates how progress is made.

The part of the brain seen in most brain maps is the cerebral cortex. It is a layer of nerve cells about 3 mm thick forming an outer shell. Something else must occupy the rest of the brain's thickness. First of all, the surface layer is infolded rather like the skin of a green pepper, except much more so. This infolding increases the total surface area of cerebral cortex many times in the human brain; some mammalian brains, such as those of the rabbit and some New World monkeys, have very little infolding.

Below the surface layer of nerve cells, there is a lot of "white matter." These are the axons of nerve cells entering and leaving the cerebral cortex. Their white appearance is due to the color of the insulation on the axons, called myelin. The white matter contains these insulated axons and the cells that support them and hold them together, called glia ("glue" in Greek). The white matter is thus analogous to a bundle of white cables running from one part of a computer to another. Gray matter is the name for the other areas, such as cerebral cortex, where insulated axons are not the overwhelming ingredient. In preserved brain, the color is grayer than that of "white matter"; in a living brain, "gray" matter is really a pinkish brown. Not only are there lots of cell bodies, dendrites, and synapses in gray matter, but there is a richer blood

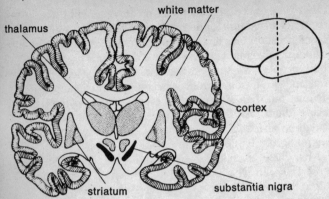

A vertical cut through both left and right brain, along the line indicated by the dashes in the side view. The gray matter of the cortex, the striatum, and the thalamus are seen. Beneath the thalamus is the substantia nigra ("black substance").

supply too, which contributes to the non-gray color.

Deep in the white matter are other collections of gray matter, the nerve cells of the striatum (also called the basal ganglia). Still deeper, where the two cerebral hemispheres join with the long core of the brain (the brain stem), lies another large cluster of nerve cells collectively called the thalamus. Beneath the thalamus is a narrow band of nerve cells whose color is much darker than the usual gray, due to a pigment derived from the neurotransmitter dopamine contained within the cells. This group of nerve cells is called the substantia nigra ("black substance" in Latin), and their axons terminate in the striatum. As we shall see, they are among the nerve cells which are destroyed by Parkinson's disease.

During the twentieth century, the electrical and chemical nature of nerve cells has become apparent. Different nerve cells specialize in the production of different neurotransmitter substances. Those in one part of the substantia nigra, for instance, happen to produce dopamine, which they release at the synapse in response to electrical signals such as impulses. In addition to transmitting one type of neurotransmitter, each nerve cell receives several different neurotransmitters from upstream nerve cells. Some of these neurotransmitters produce an excitatory effect upon the nerve cell; other neurotransmitters may antagonize, or inhibit, that cell. The push and pull is carried out by balancing the voltage of the nerve cell, the excitatory inputs raising the voltage and the inhibitory ones lowering it. Drugs may affect the process at a number of steps. For example, the amount of neurotransmitter available for release by impulses may be changed by drugs.

The second student, who has been looking through binoculars at the convolutions on the surface of Neil's exposed brain, several meters (six feet) away on the other side of the glass, turns and says, "I read somewhere about an autopsy study on the brains of Parkinson's-disease patients. I think it showed that lots of dopamine-containing neurons in the substantia nigra were missing, but no one has ever really told us the story behind Parkinson's."[1]

"All right," says the neurologist. "I'll give you the three-minute capsule summary. In 1957, Carlsson and his co-workers were experimenting on animals with a drug which caused a massive loss of dopamine from the striatum. Their animals were rigid and moved around very slowly, just like Parkinson's patients. Then in 1960 came that autopsy study that you had heard about, in which Ehringer and Hornykiewicz found that dopamine was missing from the striatum of Parkinson's patients.

"What is now obvious is that the dopamine in the striatum is the transmitter substance released from the axons of neurons which come from the substantia nigra. One of the reasons that there's so little dopamine in the striatum is that a lot of the substantia nigra cells have died. Now, you can apparently get the remaining cells to produce extra dopamine by providing them with more of the raw material from which they make dopamine—that's the theory behind giving patients massive doses of levodopa, which is the molecule from which dopamine is created. But why do the neurons in the substantia nigra die?" The neurologist poses one of her favorite teaching questions.

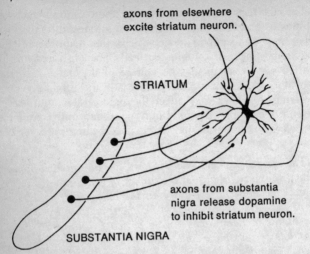

axons from elsewhere
excite striatum neuron.

STRIATUM

axons from substantia
nigra release dopamine
to inhibit striatum neuron.

SUBSTANTIA NIGRA

The group of dopamine-containing axons which travel from the substantia nigra to the striatum are responsible for inhibiting the neurons of the striatum. In Parkinson's disease, there are fewer substantia nigra neurons to inhibit the striatum.

The second student looks thoughtful. "Some cases occurred right away during that great influenza epidemic after the First World War. And lots of cases that you see today are also people who suffered a really bad case of the flu way back when. But their symptoms didn't appear for many years. If the flu killed off those substantia nigra neurons, why doesn't the disease always start at the same time as the flu attack? Why thirty to fifty years later? Does the virus hang around, slowly killing neurons as in one of those degenerative diseases?"

"A good question," replies the neurologist. "A slow virus is one possibility. But it turns out that normal people—you and I, presumably—are always losing neurons in the substantia nigra every day. It's a higher percentage than the average for the whole brain. In substantia nigra the total declines by half before seventy years of age. Presumably, the remaining neurons compensate for this loss of their neighbors. But perhaps there is a minimum number of cells that you've got to have in the substantia nigra, perhaps twenty percent of what you were born *with, or the compensation fails. Fewer than that, and you get stiff, develop the shakes, and so on."*

"So maybe the flu killed off half the neurons back in 1919, but it doesn't cause any trouble until a lot more die naturally and you reach the twenty-percent level, or whatever," exclaims the first student.

"That's the theory. Or at least, that's one of the theories floating around, waiting to be tested. But essentially, there's a group of neurons which are overactive because many of the neurons which inhibit them have died."

"Why not compensate by just shutting down some of the excitatory inputs which increase activity in the striatum? If there's less inhibition, why not less excitation? Wouldn't that balance things?" asks the first student.

"Good idea. Actually, there are opportunities all along the chain of neurons because, at each neuron, you have both excitatory and inhibitory inputs. Some push backward and

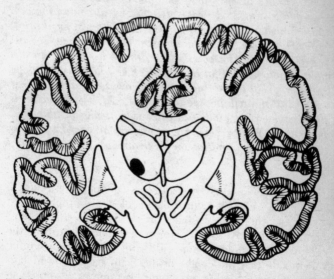

A thalamotomy for Parkinson's disease destroys approximately the area shown in black.

others pull. Presumably, that's how the surgical treatments for Parkinson's disease have their effect. Neurons deep in the thalamus are destroyed in such operations, either by overheating them or by freezing them. We're not sure exactly what role those neurons play, but there's good evidence that they inhibit other neurons in the chain of neurons controlling muscle tone, so their loss helps restore the balance between push and pull. I think you said earlier that this patient had such an operation."

"Yes, and he was aphasic for a while afterward. It was done in left brain."

"Any changes in his motor function after that operation?"

"Well, he said that he used to shake more in the right arm and it was much stiffer than the left. But in the ten years since that operation, the right arm doesn't shake at all. The left arm tremor has gotten worse, though, and his face more masklike. Does anyone do operations like that anymore?"

"In the decade before the discovery of levodopa as a treatment of Parkinson's disease, that thalamic operation was the best available treatment. It was quite effective at controlling arm and leg tremor and stiffness, but didn't do much for the slow movements, voice fading, face movement, swallowing, or gait. So we still use it for cases where arm or leg tremor are the main symptoms. But for those with other symptoms, levodopa is better. That's how medicine progresses, by adding new treatments and refining the uses of old ones. The operation is also useful for some other diseases with tremor or abnormal movements. By the way, you never explained why he was aphasic after the operation on his thalamus."

Traditionally, language has been considered a function only of the cortex. It seemed reasonable that so sophisticated a function should be in that part of the brain that is most highly developed in man. However, about the turn of the century, a neurologist named Pierre Marie argued quite vehemently that the depths of the brain had a role in language. Despite Marie's arguments, it came as a surprise to most investigators that surgical lesions in the left human thalamus sometimes led to a language disturbance. Usually this was a temporary difficulty in naming things, but occasionally it was permanent. Equally surprising were observations of altered language after small hemorrhages confined to the left thalamus. Such observations were made only after the development of diagnostic techniques that allowed the identification of this type of damage during life. Large thalamic hemorrhages destroy so many structures that patients are in coma, their language function untestable.

Smaller hemorrhages weren't regularly diagnosed until techniques like the computerized tomographic (CT) scan came along. CT scan clearly shows small thalamic hemorrhages and, in those cases with left thalamic hemorrhages, a new pattern of language disturbance has been identified: there is fluent speech (as in Wernicke's-area cortical damage), but the same word is repeated over and over (as in Broca's-area cortical damage), and there are wide swings in performance, from nearly normal to unintelligible. Wildly inappropriate words appear in these patients' conversations. For example, one patient with a small left thalamic hemorrhage repeatedly used the phrase "affirmative action" for names of simple objects. [2]

These left thalamic areas concerned with language can be identified during the thalamic operations for Parkinsonism (thalamotomy) by stimulation mapping during naming, just as cortical areas have been in Neil's operation. Such studies suggest that there is a common mechanism in the thalamus for both language and memory. That common mechanism focuses attention on those things in the environment that can be labeled with words. When this mechanism is damaged, attention cannot be held on anything long enough to produce an appropriate name. Instead, attention immediately shifts and a random extraneous word pops out. Like the left cortex, only the left thalamus is concerned with words. The right thalamus has an attention mechanism for shapes. In the thalamus the attention mechanism for words is clearly dominant. When it is turned on, shapes are ignored. But turning on the attention mechanism for shapes doesn't change the use of words. [3]

This thalamic attention mechanism can be manipulated by

electrical stimulation, with most unusual results. If the stimulating current is applied to the thalamus when information is coming in, that information will be remembered, seconds to days later, about twice as accurately as similar information coming in without stimulation. If the same current is applied when information already in memory is to be retrieved, retrieval is faster than usual but with many more errors. This attention mechanism seems to regulate what comes into or out of memory at any given moment, almost as if it were opening or closing a gate. When something is coming in, similar information cannot simultaneously be retrieved from memory. The attention mechanism also determines how easily something coming in will later be remembered. Most inputs have both verbal (word label) and spatial features. The relative strength of this attentional mechanism may determine how many of those verbal or spatial features are remembered, and the number and types of associations available from memory to relate to this new input.

Enhancing memory for incoming information by thalamic stimulation may eventually become an aid in rapidly retraining stroke patients with language disorders. One can imagine science-fiction uses, too—for example, helping medical students to study for exams.

Thalamic stimulation produces effects on short-term memory. Top block shows a three-slide sequence with stimulation during object presentation; such stimulation *decreases* errors during recall (there is a 36 percent error rate in the OR without stimulation). Middle block shows stimulation only during the recall attempt; errors *increase*. Lower block shows the control experiment, where stimulation occurs during both presentation and recall; errors are merely average (data from G. Ojemann 1975). This thalamic-stimulation effect is thought to represent activation of a system that directs attention to verbal material in the external environment (hence the decreased errors when stimulation was used during presentation), while blocking retrieval of verbal material already in memory (hence the increased errors when stimulation occurs during recall attempts).

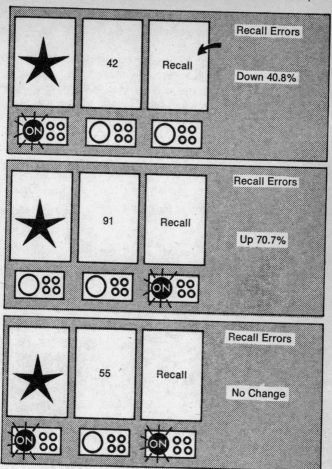

"But doesn't this complication of aphasia make the thalamic operation too risky as a treatment for Parkinson's disease?" one of the students asks.

"Fortunately, the aphasia is usually transient, disappearing in a few days," the neurologist remarks as she turns to prepare her EEG machine. "But there is a small risk of persisting aphasia and other neurologic deficits with that operation. And, for that matter, with the one going on here today—a few chances in a hundred in each case. But there are risks to all treatments, you know, whether drugs or surgery. Levodopa carries risks, too.[4] Medicine is always a process of balancing off these risks against the chances of helping. No operation or drug is risk-free, though Congress and the Food and Drug Administration seem to forget that sometimes."

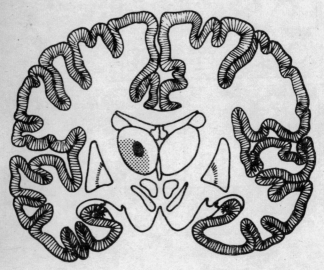

Area of thalamus where stimulation changes memory for words (dots) and where stimulation changes language performance (black).

6

Learning and Remembering: How Are Memories Recorded?

Studies of the psychology of memory in man and other animals indicate that memory includes several processes:

1. An *immediate memory*, usually tested by reading someone a long string of numbers and asking them to repeat it back. Most people can remember a string of about seven digits, e.g., a telephone number. In such a test, all of a person's attention can be directed toward the task of hanging onto the numbers.

2. A *short-term memory*, also called post-distractional memory because the subjects are not allowed to devote all of their attention to the memory task. As with Neil's task in the OR, another engrossing task (such as counting backward by threes) is used to keep the subject from rehearsing the correct answer. An everyday example of a short-term post-distractional memory task is to try to recall what you had for breakfast.

3. A *long-term memory*, used to remember your name, the name of the President of the United States, your childhood, etc.

There are good reasons for believing that different mechanisms in the brain are used for short- and long-term memory storage. Short-term memory can be disrupted by something which interrupts brain function: a blow on the head, an epileptic seizure, or a brief interruption of the brain's blood supply. Immediately after recovery, the person may have trouble remembering what happened for the minutes (or hours, sometimes even days) prior to the brain disruption. This amnesia does not, however, extend to long-term memories. Even deep coma, which may silence much of the electrical activity of the brain, does not erase long-term memories.

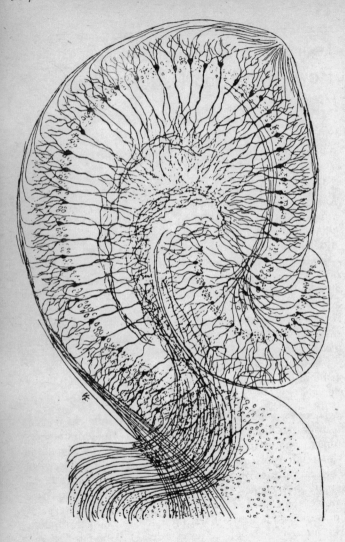

Layering of a simple cortex: the hippocampus and the dentate gyrus (Krieg 1957).

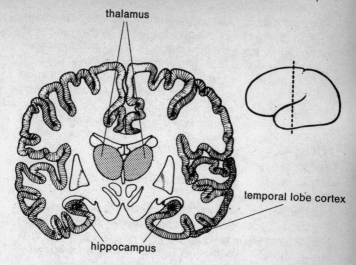

Parts of the brain involved in memory are dotted.

Long-term memories are not formed immediately after a new piece of information is acquired. New memories are "held" in short-term memory; animal experiments suggest that the more permanent long-term memory is formed gradually over a period of many hours or days. Disruption of brain function means not only that the memory is lost from the short-term memory mechanism but also that the permanent memory trace is imperfectly formed.

Areas of brain involved with the short-term memory mechanisms have been partially identified. Regions of language cortex play a role, as noted in describing Neil's inability to recall a name following stimulation during the distraction. Portions of human thalamus seem to help focus attention, gating access into short-term memory and changing the "intensity" with which incoming information will be remembered. And many studies in animals and man have indicated that a structure on the inner side of the temporal lobe, the hippocampus, is centrally involved in the short-term memory process.

The destruction of the hippocampus on only one side of the brain produces a minor memory deficit in man, so minor that careful neuropsychological testing is required to demonstrate it.[1] The deficit is for handling verbal material if the left hippocampus is damaged; for the right side, the losses involve small changes in the ability to remember new faces, other shapes, and music sequences. These deficits do not significantly alter the quality of a person's life. For example, part of the left hippocampus will be removed in a patient such as Neil. The surgery thus involves something of a tradeoff. For Neil the chance to eliminate or control his seizures is worth the minor changes in verbal short-term memory abilities which will probably result.

Destruction of both right and left hippocampus produces far greater deficits than might be predicted from the results noted above. Many years ago, a neurosurgeon operated on a patient who had epileptic foci in both right and left temporal lobes and destroyed both right and left hippocampus.

H.M. is now forty-seven years old. He cannot recognize his next-door neighbors, because his family moved to the present house twenty years ago, shortly after his operation. Once, when a psychologist who was driving him back from a medical center asked him to help her find the house, he guided her to a street that he said was quite familiar to him, though he admitted it was not the right address. The psychologist then called his mother and learned they were on the street where H.M. had lived as a child, before his operation. During his stay at the medical center, he had kept ringing for the night nurse, and with repeated apologies asked her over and over again where he was and how he came to be there. His doctors noticed that if they walked out of the room for even a few minutes and then returned, he did not know them—they had to be introduced to him all over again. Now H.M. sits at home, alone with his aged mother, next to old issues of the Reader's Digest, which seem eternally new to him. Everything he reads vanishes from his mind, as if a slate had been wiped clean. (From Maya Pines, The Brain Changers, reprinted with permission of Harcourt, Brace, Jovanovich, Inc.)

H.M.'s immediate memory is normal, as is his long-term memory for events more than several years before the operation.[2] The problem is with his short-term memory and thus his ability to form new long-term memories.

The neurosurgeon immediately publicized this result so as to prevent others from suffering a similar fate. H.M. is not, however, unique in suffering this deficit in short-term post-distractional memory and the resulting inability to form new long-term memories: there are patients whose strokes have destroyed the hippocampus on both sides. A similar condition, called Korsakoff's psychosis, occurs with damage to the inner portion of the thalamus. This is typically a disease of alcoholics, thought to be due to the lack of Vitamin B_1 in their alcoholic diets (it has been proposed that alcoholic beverages be "fortified" with B_1, in the manner of milk, to

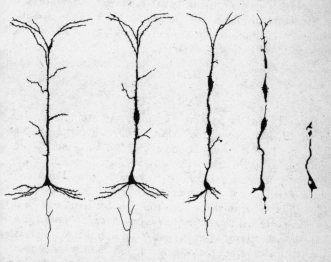

Sequence of dendritic tree changes in senile dementia (from Scheibel and Scheibel 1977).

prevent the disease). These patients can often be discovered making up stories (confabulating) to cover up for their inability to remember.

In contrast to these various brain areas which have been identified with the short-term memory process, it has been quite difficult to associate any particular brain structure with long-term memories. There seems to be no one area of brain where damage will leave a patient with the loss of long-term memory. The closest examples come from the patient with damage to the brain stem, who may go through a period in which he is awake and talking but cannot recall such simple long-term memories as his name and birthday. This, however, is apparently a failure of the retrieval mechanism, not the loss of stored information; the patient may remember everything later in recovery (except for being amnesic for the period just before and following the injury). Severe dementia, where many neurons in many cortical areas die, also produces a long-term memory deficit.

Several decades ago, the Montreal neurosurgeon Wilder Penfield described some patients in whom electrical stimulation of the temporal lobe seemed to be able to "tap into" specific long-term memories. These observations were made during epilepsy operations such as Neil's.[3]

Point 23, in the left temporal lobe, was stimulated. The Montreal neusurgeon asked Maria if she felt anything. She told him about the song she had heard. But she couldn't quite place the song. Perhaps it was a baby song. Stimulation was repeated: she heard the same song again. It was "War March of the Priests," not a baby song after all. The surgeon said, "You didn't hear anything, did you?" as he stimulated point 23 again. He was wrong; she had heard that same song again. Now she remembered where she had heard it; it was the back side of a particular record that she used to like to play many years ago. "The Hallelujah Chorus" was on the front side. She remembered how much she had liked that record.

Only a few patients reported such flashes of memory upon electrical stimulation; they often were those patients whose temporal-lobe seizures also evoked brief flashes of previous experience. One possibility is that the long history of seizures in these patients made some memories more accessible, so that even local electrical stimulation sufficed to evoke them.

Some patients' memories were auditory (especially for left temporal lobe stimulation) and some were visual (especially

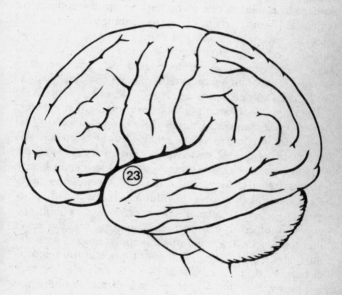

The memory of a specific childhood experience was evoked from this site in Maria's temporal lobe; she heard music ("The War March of the Priests").

for the right side). All had the quality that everything about a specific event was recalled, even down to the fine details and the emotional coloring. These findings led Penfield to propose that we have recorded all of our conscious experience (or at least all to which we were paying attention) somewhere in the brain, with the temporal lobe assisting in gaining access to that record.

Just because we can no longer recall something does not mean that it is lost, no longer recorded somewhere in the brain. The earliest studies on the psychology of memory were on people learning a foreign language. People who learn a language in early childhood, but who then switch to another language, often lose their ability to understand the early language. Yet when they go to relearn the language, they gain proficiency much more rapidly than others. This "savings" can be demonstrated for many kinds of learning.

The degree of retention of a piece of information probably depends on a complex of things that occurred upon original acquisition: how intensely one attended to the material, and how much emotional response was stirred up. During the hours-to-days transition between short- and long-term storage processes, other factors are also likely to operate: how many associations did that piece of information make with other material already in memory, how much competition was there with other similar incoming information (the fewer things happening, the better might be the storage of what did occur), and the like. The number of times that the information is recalled and used affects the ease with which it may again be recalled. Finally, at the time of the recall attempt itself, how many associated items are already at the level of consciousness (perhaps in short-term memory again because of having been recalled minutes or hours previously)? Each of these processes probably affects the success of the recall attempt, each probably occurs using a different combination of brain areas, and each can probably be manipulated to improve long-term memory.

The foregoing discussion about memory indicates just how thin our understanding is concerning one of the most important aspects of the brain. There has been much speculation

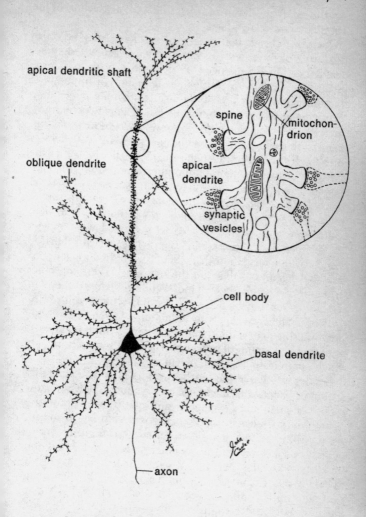

An idealized drawing of a large neuron in visual cortex. Inputs from other neurons usually occur at synapses upon the tiny "spines" populating the dendrites. The enlargement indicates how a dendrite looks at higher magnification (from Greenough, 1975).

about memory mechanisms; what follows is a selection of the hypotheses and related facts.

One possibility is that long-term memories may be coded in a string of molecules, in the manner of genetic memory, which uses DNA sequences. This would explain why long-term memories are so resistant to temporary disruptions of brain function; they are "built in" to neurons rather than depending on minute-to-minute electrical activity of neurons. But the DNA sequence is changed only by random mutation, not experience; experience might change RNA sequences or proteins. If this is the case, one might be able to extract the relevant chemical containing the information from the brain of a smart animal and give it to a dumb animal to make it smart. In the 1960s, there were reports of exactly this sort of result: worms trained to respond in a certain way to light flashes were ground up and fed to untrained worms, who then learned the light response faster than expected.[4] This led to the obvious suggestion that students would no longer need to attend lectures if they ate the professor. There are other difficulties with this theory, too. While protein synthesis in cerebral cortex may be necessary if a long-term memory trace is to be established (applying inhibitors of protein synthesis after exposure to new information interferes with its retention), there are many technical issues associated with such experiments which complicate their interpretation.

Besides changes at the biochemical level, it is thought that anatomical alterations might occur in neurons following learning.[5] The anatomical change is unlikely to involve just one neuron or synapse: with aging, neurons are lost but memories are not similarly lost in a piecemeal fashion. A given neuron probably cooperates with many other neurons in forming the trace of a particular memory. As with people trying to push a car out of a ditch, no one person may be essential; there may be a minimum number of people needed, and it is likely that the task will be accomplished more efficiently if everyone works cooperatively. A given memory may involve changes in many neurons, perhaps involving a number of different specialized regions of the brain.

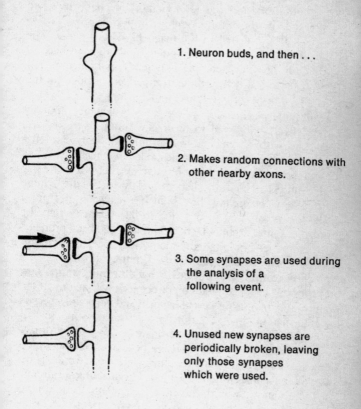

1. Neuron buds, and then . . .

2. Makes random connections with other nearby axons.

3. Some synapses are used during the analysis of a following event.

4. Unused new synapses are periodically broken, leaving only those synapses which were used.

The "photographic development" model for the long-term memory trace.

Our thinking about possible memory mechanisms is strongly influenced by the analogies which science and technology have provided us thus far: the term "laying down the memory trace" probably originated by analogy with a groove being cut on a phonograph disk. Then came the digital computer having both short- and long-term memory mechanisms: when the power fails, the information in the transistorized active registers is lost, but that in the magnetic core survives. The analogy is defective in that computers store a given piece of information in a discrete pigeonhole. The hologram has provided us with a better, though imperfect, analogy for the distributed aspects of long-term memory storage.

Evolutionary principles, where random genetic variations are selectively conserved to form new animal species, also provide us with an analogy for the formation of new long-term memories by conserving some random connections. While the number of neurons does not increase after birth, the size of the individual neuron changes greatly. In infancy, some neurons send out many short branches (called spines) for a while; this is followed by a period in late infancy when the number of spines declines by half, as if they had been pruned.[6] Another model for long-term memory storage in the adult brain has been based upon similar pruning. It assumes that even adult neurons are continually budding and sending new branches out to make random connections with axons in the vicinity. When some new event happens and the brain processes the information, many neurons are activated. Suppose that the new synapse happens to participate in this process. Imagine a pruning process that breaks new connections which were not utilized in the preceding day. This would leave intact those synapses which were used. The anatomical substrate for the memory would therefore take some time to be "developed," as in the photographic development of a negative where unexposed silver grains are removed. Thus, laying down the permanent memory trace might simply be a matter of conserving those randomly made connections which happen to participate in responding to a new event. Short-term memory might involve the temporary strengthening of a synapse (see facilitation in Chapter 10) following extensive use by increasing for a day the amount of neurotransmitter released by an impulse. Indeed, the facilitation changes in the synapse might be what protects the new synapse from the pruning process. Such a theoretical model helps to explain why, although short-term memories are easily disrupted by massive activation as in seizures (analogous to fogging the film), long-term memories could require some time to develop and be disruption-proof thereafter.

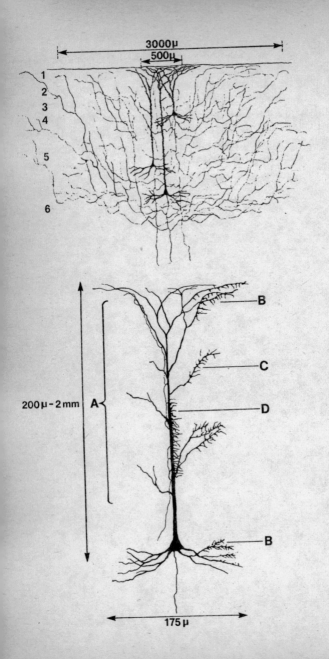

7

Left Brain, Right Brain: Shapes, Words, Art, and Music

With speech and language has come the development of some notable differences between the left and right sides of the brain. In most other animals, the two halves are mirror images of each other. They not only look the same, but the right brain does for the left body what the left brain does for the right body. Not so in man, for speech and language are usually only on one side—the left in most people.

Knowing on which side language is located is obviously crucial for the neurosurgeon, since this is an area he must certainly avoid damaging. In Neil's case, the side of his brain which housed language was determined during a test done several weeks before the operation. In that test, a short-acting barbiturate drug was injected into the artery that supplies one side of the brain, temporarily putting to sleep just that one side. This was done while the patient was busy naming the objects shown by a slide projector. After one side of the brain was briefly put to sleep, there was a rest period and then the other side of the brain was similarly put to sleep. The side anesthetized when the patient stopped naming objects was the side housing the language portions of the brain.

An easier but less reliable method of telling which side houses language is to use a "dichotic listening test": two

Synaptic inputs to a large cortical neuron come from many sources. Those from the reticular activating system (A) are widely distributed up the dendritic tree. Inputs from neighboring cells (B) arrive at the top and bottom dendritic branches. Inputs from the opposite hemisphere (C) and the thalamus (D) end upon middle regions. At top: Drawing of three large cortical neurons shows their axons entering the white matter at bottom, but axon branches going to neighboring regions of cortex. Note that axons spread over a wider area than do dendrites (from Scheibel and Scheibel 1970).

words differing from one another by only one sound are simultaneously presented to the two ears. After several sets of these word pairs, the patient repeats back the ones he heard. The ones he remembers are usually those presented to the ear opposite to the side of the brain housing language; thus, if "peace" is presented to the right ear at the same that "cease" is presented to the left ear, the patient will tend to report "peace" if he is left-brain-dominant for language. This kind of test works for reading too (and is called the "dichhaptic test"). The left brain initially gets all of the visual input from the right half of the visual field, and vice versa. If words are briefly flashed on the screen to the left and right of a point on which the patient is fixating, the word on the right side will be more easily recalled by a patient whose left brain is language-dominant.

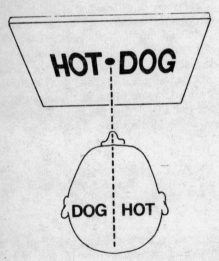

Dichhaptic testing setup. Subject looks at dot in center of screen and then words are briefly flashed onto the screen to the right or left of this dot. For example, a subject with language in the left brain will tend to report only the "DOG" part of "HOT DOG" if the flash is brief enough; the right-visual-field material is initially processed by the left brain's visual cortex, which is closer to the (left) language area than is the right visual cortex.

These and similar tests indicate that all language output comes from left brain in most people. The left side is also heavily involved with the ability to read, manipulate grammar, and regulate the sequence of movements of the face and hands. The right brain, in most people, has the ability to manipulate things in space, follow maps, remember shapes and faces and musical tone sequences, relate clothes to body image—it even seems to house the body image itself. Thus, the "average" human has a verbal left brain and a visual-spatial right brain.

But these abilities are not "lateralized" in the same way in everyone. Left-brain damage disturbs language in about thirteen people for every one person with language disturbance after right-brain damage. The ability to manipulate clothing in relation to one's body, as in getting dressed, is altered by right-brain damage in five people for every one person with dressing problems after left-brain damage. Right-brain damage interferes with the ability to manipulate objects in space, or follow routes, only about twice as often as left-brain damage. So it is not simply a matter of language on one side and visual-spatial on the other side—the mix is highly variable. Some individuals have both language and some (or all) visual-spatial functions crowded onto the same side. But we do not know the consequences of this for function, except in the most extreme cases.

The degree of lateralization depends in part upon one's sex. Males are more likely to have visual-spatial abilities strongly lateralized to the right brain than are females. Failure to develop this strong lateralization of visual-spatial abilities has been identified in some boys with developmental dyslexia, a condition in which an otherwise intelligent child has great difficulty in learning to read. Dyslexia has a genetic component and is most often seen in males. These dyslexic boys have language in the left brain, like most people, but have visual-spatial abilities on both sides, unlike most males. This raises the possibility that the left language areas are squeezed by the left visual-spatial areas, to the detriment of reading.[1]

There is a rough correspondence between handedness and this "lateralization" of language. Most, but not quite all, right-handers have language in the left brain. Left-handers, who are 15 percent of the population, are another matter. More than half of them also have language on the left, bringing the total of left-brain language up to 93 percent of the population. About half of the remaining left-handers (3 percent of the population) are true mirror-image reversals, with language in the right brain.[2]

The remaining left-handers have language on both sides of the brain. This particular mix seems to be a less than optimal arrangement, since these people seem prone to language disorders, particularly stuttering. There are cases in which a stroke, involving only one side of the brain, has cured a lifelong stuttering problem, suggesting that stuttering occurs when the two sides are both involved with language and have difficulty in coordinating their act.[3]

The very young show less evidence of lateralized functions; indeed, they seem to have the ability to move functions around following brain injury.[4] The earlier the damage occurs, however, the better the chances for this compensation to occur; after about the age of five, such flexibility is lost. For example, there is a congenital disorder in which abnormal blood vessels develop on just one side of the brain. If this condition is left untreated, severe seizures occur which not only affect the abnormal side but spread to the normal side and interfere with its development. To save the good half of the brain, neurosurgeons remove the abnormal hemisphere; the operation usually is done before six months of age.

If it was the right hemisphere which was removed, language develops normally. In those who lost left hemispheres,

Spotting the anatomical differences between left and right halves of the brain requires a view of the top of the temporal lobe, achieved by cutting the brain below the line shown at top in the side view. Looking down upon the cut brain (bottom), the infoldings of the rear portion of the temporal lobe are seen to differ between left and right sides (arrows). The dotted area is called the "planum temporale" and is much bigger (in most people) in the left brain. The function of this area is not well established, as stimulation mapping is usually prevented by its buried location, but the large left planum temporale is adjacent to the language cortex. The auditory-receiving area (see figure on page 9) is just in front of the planum temporale.

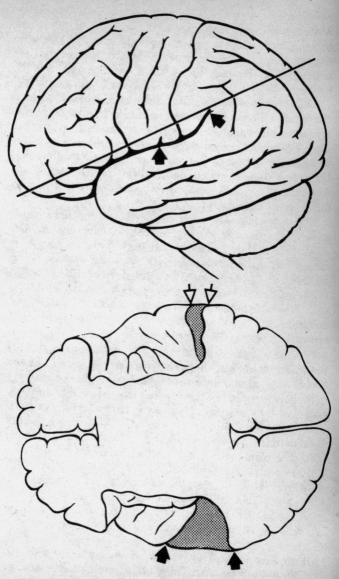

speech develops, but not fully. They are considered quiet children, who use language only when necessary and then with a reduced range of grammatical expression. They tend to talk in the present tense; more elaborate constructions are beyond them. For example, when asked to repeat back the sentence "Wasn't the poor cousin helped by the old lady," one nine-year-old who had had the left hemisphere removed shortly after birth said, "Wasn't by the cousin helped by the old lady," while another recalled this sentence as "Wasn't the poor cousin . . . helped the old lady." Children with right-hemisphere removals make few such grammatical errors. Apparently there is something wired into only the left brain, even at birth, that is essential for the full range of human grammatical expression.[5]

When language has been displaced into the right hemisphere by a missing left hemisphere, visual-spatial abilities also suffer, usually more than language. This, in a sense, is the reverse of the situation in dyslexia: in the dyslexic child, the abnormal presence of visual-spatial functions in the left brain impairs language; in these right-brain-only children, the abnormal presence of language impairs visual-spatial functions. Perhaps it requires more brain than is available in one hemisphere to do both things well.

Despite this ability of the other side of the brain to pick up the load in early life, the underlying basis for lateralization is probably wired into the human brain. A number of anatomical differences can be seen between left and right brain if one looks carefully enough.[6] These asymmetries are present in the developing fetus, not just in the adult. The best-known of these asymmetries involves a region on the top of the temporal lobe (the planum temporale), which is about the size of a large coin on the left side but the size of a small coin on the right side. This asymmetry is reversed (right side bigger than left) in just about the proportion of people that one would expect not to have language in the left brain. Like lateralization of visual-spatial function to the right brain, this asymmetry is greater in men than women.[7]

If this anatomical asymmetry is concomitant with the lateralization of language, might other animals also have it? Might it mark some of the final evolutionary steps toward language? Such an asymmetry has been identified in chimpanzee and orangutan, although their asymmetries are far less extensive (and thus harder to spot) than those of man. Chimpanzees have been taught simple language (using sign language, since chimps lack the requisite vocal development for versatile speech).

The right brain is concerned with visual-spatial functions. Strokes that damage the right brain may leave the patient looking remarkably normal except, perhaps, for left-body paralysis if the stroke is large enough to also damage the right motor strip. But the fluent speech of such stroke patients conceals a severe deficit, a deficit that could have major consequences.[8] Imagine the following drama:

The President of the United States is lying in bed, waving his right hand at his Secretary of State in a gesture of dismissal. The President is alert and seems intelligent. He is talking forcefully, angry at his subordinate, who has suggested that the President is ill and perhaps should delegate some of his duties to others until he recovers.

Indeed, the President's left side seems to be totally paralyzed from a recent stroke. His left arm lies limp. The President cannot walk because his left leg will not function. However, the President seems blissfully unaware of this disorder, steadfastly denying that there is anything wrong with him. It is, of course, this denial of his illness that has particularly upset the President's personal and official families. They have tried to reason with him, pointing out that his left arm is lying there, paralyzed. But he denies that it is his left arm. Indeed, he is somewhat puzzled about what a strange arm and leg are doing in his bed with him.

President Woodrow Wilson suffered a stroke in 1919, after he triumphantly negotiated the League of Nations charter but before he attempted to persuade the U.S. Senate to ratify it. The stroke paralyzed his left side. He denied his illness to the point of paralyzing his administration; he fired his Secretary of State for discussing the illness with the Cabinet. Such effects of right-brain strokes were only beginning to be recognized by neurologists of that era. Today, the "denial of illness" syndrome is well known, and the typical symptoms of its victims have been ascribed to our president of the above paragraph. It illustrates how the right parietal lobe plays a role in the perception of one's own body image; if one cannot perceive anything about the left side of the body, to say nothing is wrong with it does exhibit a certain logic.

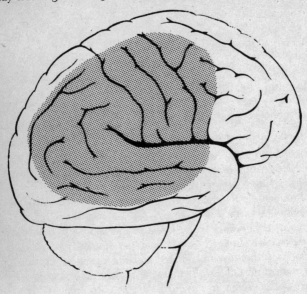

The "President's stroke," a typical massive stroke caused by the blockage of the right middle cerebral artery.

Less severe strokes provide another illustration of parietal-lobe function. In a small right-brain stroke confined to the parietal lobe, there may be no paralysis, no visual defects, no other sensory defects of the usual sorts. Indeed, the person appears normal. But if the patient is driving a car approaching an intersection, though he can see cars approaching from the left just as easily as he can see cars coming from the right, if cars approach from both directions at the same time, the patient is likely to pay attention only to the one coming from the right. He may never realize that the left one exists, unless, of course, a collision happens to occur. If touched on both hands at the same time, he may notice only the touch to the right hand.

It may be very difficult to persuade such a patient that he should not drive a car. He may not be able to tell that anything is wrong with himself. He cannot perceive his own deficit in perception. It is this loss of insight into one's own abilities that is a particularly devastating feature of these strokes; these are among the most difficult brain-damaged patients to rehabilitate. Such unawareness of a stroke is an exception to the general rule. Disorders of the brain are truly dehumanizing conditions, made all the more unbearable because of the patient's often agonizing awareness that he or she is no longer the same person.

Damage to the right parietal lobe can also produce deficits in other visual-spatial abilities. The patient may have difficulty reading a map, or finding the way from bed to bathroom. He may have difficulty dressing himself. He can move all of his muscles in a coordinated manner but his arm may wind up in a pants leg rather than in a sleeve, even though he can name sleeves and pant legs and describe what they are used for. As with our President, such patients seem to have a disturbance in their internal image of their own bodies.

If the patient is an artist, his drawings may be distorted, perhaps only filling in one side of the canvas, or only developing one side of a face. Neurologists regularly test forms of artistic skills as a way of detecting right-parietal-lobe damage. Such patients produce defective drawings, even if only copying a cross or a house or a clock face. Features are omitted and others are crowded over onto the right side of the drawing. A clock face may have all twelve numbers, but crammed into the right side between twelve and six o'clock; the left-side positions between seven and eleven are ignored. The patients seem to realize that clocks should have twelve numbers but cannot space them out around the left half of the circle.

These deficits not only affect the performance of an artist, but many other skills as well. The auto mechanic can identify and explain the function of all the parts of an engine after such brain damage, but is unable to assemble the parts. It has been suggested that unusually good functioning of this nondominant (usually right) parietal lobe is a feature of those who excel in the visual arts: painters, sculptors, architects, movie makers.[9]

Other skills seem to depend on function of both sides of

the brain. Mathematical calculations are often impaired by damage to either parietal lobe, with division and subtraction more readily impaired than addition or multiplication. Dichhaptic testing suggests that the right brain is more important than the left for this calculating ability. The defect with left parietal damage may, in part, be difficulty in labeling mathematical symbols with words, for it is sometimes associated with several other symptoms that involve putting word labels on "spatial" material: inability to distinguish left and right and to name body parts (particularly, and sometimes only, the inability to name the fingers). Writing is also frequently impaired, as is spelling; yet, in some such patients, other language abilities are largely intact, such as naming or reading.

Musical abilities also seem to be distributed on both sides of brain. The right brain plays a major role in music, for during the time that only the right brain is put to sleep by barbiturate injections, the patient sings in a monotone, having lost the ability to correctly reproduce pitch; rhythm is less disturbed. Temporal lobe seems to be especially important to

Drawings following a small right-parietal-lobe stroke, damaging the area shown in the right-brain map.
1. A cross drawn by the examiner for the patient to copy
2. The patient's attempt at copying the cross
3. The examiner's drawing of a house
4. The patient's attempt at copying it; again, the features on the left side are omitted
5. The examiner drew a circle, added some hair, and asked the patient to draw a face inside the circle; the patient drew features only on the right side of the circle (he said that the tongue was sticking out).
6. The examiner drew the circle indicated by the arrow and asked the patient to draw a sunflower; the patient filled in the petals and leaves

After all of the drawings were complete, the examiner asked the patient if anything was wrong with any of the drawings; the patient said no. The patient could name the object in each drawing, could name the various parts of the face and sunflower, and was apparently quite oblivious to the missing features in the left halves.

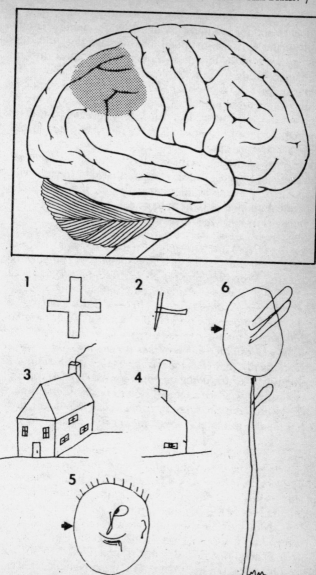

music. Right-temporal-lobe damage disturbs memory for tone sequences and for loudness, though rhythm memory is relatively unaffected. Left-temporal-lobe damage does not alter memory for tone or loudness, though it may affect rhythm. Yet damage to parts of the left temporal lobe will sometimes greatly interfere with musical abilities. Emotional responses to music are said to be diminished after left-brain damage. It has been suggested that skilled musicians develop a left-brain dominance for music compared to the right-brain dominance of the average person.

The areas of the left brain involved in music are separate from those for language, however, as illustrated by the following two patients. Because of this, music therapy has been useful in the rehabilitation of some aphasic patients, especially those with deficits in language fluency.

Ron and Norman were admitted on the same day with blood clots in their left brains. Ron's hobby was singing folk songs. After recovering from the acute effect of his hemorrhage, he found that he couldn't sing at all. Neither words nor tune came, even for his old favorites. His speech was pretty good, though, just an occasional name that wouldn't come to mind. In contrast, Norman's speech was terrible, full of jargon, and sometimes he couldn't follow verbal commands. But a speech therapist finally got across the idea of singing to Norman, and he sang very well, including all kinds of words that he couldn't speak correctly. CT scan showed Ron's and Norman's hemorrhages to be about the same size, but Ron's was in the front part of the left temporal lobe, Norman's in the back part.

As far as the brain is concerned, there seems to be a difference between speaking and singing the same string of words. Does the tune provide a framework which helps Norman put the words in sequence? Does Ron's inability to sing reflect a failure in the coordination between word strings and the tone sequence? We would like answers to such speculative questions, but how can we get them? In science, one tries to phrase the question (i.e., design the experiment) in such a way as to force nature to give an unambiguous answer. Strokes are intrinsically messy experiments performed by chance. Electrical stimulation, as we have seen, is more localizable and controllable (and, most important, reversible!).

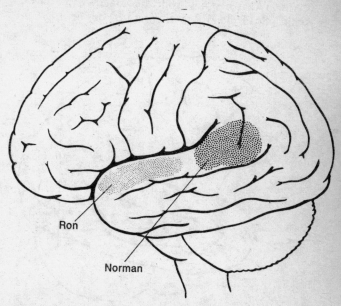

Speech vs. music: location of hemorrhages in Ron and Norman.

But the brain of an epileptic may not be normal (in the sense that many epileptics have suffered seizures since childhood, and the brain's flexibility during development might rearrange maps); indeed, this is a major problem in interpreting split-brain studies to infer the capabilities of the right brain. Another important aspect of experimental design is what one asks the patient to do. One must accurately measure a particular, limited kind of behavior (such as naming, post-distractional memory for words, phoneme recognition, etc.). If one wanted to design experiments around our speculative questions relating word sequence to tone sequence, one would have to come up with some ingenious verbal tests and then find some technological method of measuring or disrupting neuronal function.

Thus, the "maps" of cortical function may be biased by the techniques available and by the questions asked. But there is also another way of finding where different brain functions are located. When neurons work, they increase the blood flow locally. So if blood flow is measured during a test of

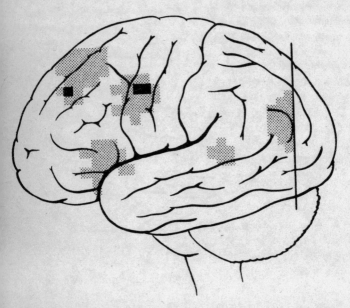

Silent reading increases the blood flow in certain areas of the brain (shaded areas had moderate increase in blood flow; black areas had large increases). Much activity would also be expected in the rear (visual cortex), but the technique cannot measure flow to the rear of the vertical line (adapted from Lassen et al. 1978).

specific behavior such as reading, the areas of brain that are working harder during that behavior can be detected.[10] Such blood-flow measurements can be made through the intact head by introducing a radioactive tracer in the blood (radioactive xenon is commonly used) and placing a series of radioactivity counters over the scalp. The counters are designed to measure just the radioactivity coming from brain directly beneath them. The rate at which the tracer is cleared from beneath the counters is determined, a measurement proportional to blood flow. At present this technique cannot adequately discriminate between blood flow in cortex and that in underlying brain. But such measurements confirm, in the intact person, many of the findings derived from the study of stroke patients or from stimulation mapping of epileptics.

Each of the techniques has its limitations, but together they give us a picture of cortical localization. The blood-flow techniques have also allowed studies to be made of the cortical areas active in mentally rehearsing an act without making any movements. In silent reading, parts of the frontal lobe become quite active, together with language and visual areas; when reading aloud, the motor strip's oral and facial areas also increase their activity. This may suggest an orchestra with the conductor bringing in the trumpets to add to the theme developed by the strings and the woodwinds. But to what extent is this analogy valid, or is the orchestra conductor really just another version of the "little-man-in-the-head" reasoning which has plagued our thinking about the brain? To explore the topic, we will consider the brain's mechanisms controlling attention and arousal, initially by illustrating some of their malfunctions.

HIPPOCAMPUS

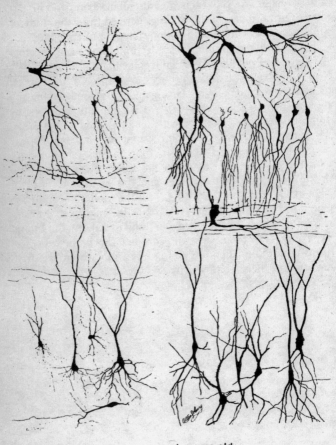

human at birth   six years old

8

Seizures:
Electrical Storms of the Brain

Neil is moving around restlessly, not an uncommon problem after four hours on the operating table, but the anesthesiologist reports that Neil is making fumbling movements with both hands.

"Neil, how are you feeling?" asks the neurosurgeon.

"Feel funny. . . . I see that man walking his dog again. . . . He's getting bigger, and he's almost to the middle. . . ."

"Neil? What's happening now?"

There is no response from Neil. The fumbling movements of his hands become more pronounced, and he tries to move around but is restrained by the anesthesiologist and the circulating nurse.

After a minute, Neil quiets down but still does not respond to questions.

The neurosurgeon takes the opportunity to describe what is happening to those medical students watching from the gallery.

"What you just saw was one of Neil's typical seizures. He often has this hallucination, the man walking his dog. That's the 'aura' which heralds his seizures. He knows that when the man and dog reach the middle of the picture, he is going to lose consciousness and not remember anything afterward. It's called a psychomotor seizure, and it only occasionally progresses to become the typical major motor seizure, with all of the rhythmic twitching of arms and legs."

"Neil, can you hear me?" the neurosurgeon asks repeatedly for several minutes.

"Wha . . . What'd you say?"

"Neil, tell me what happened."

"Don't know . . . guess I had a seizure, didn't I?"

"Did you hear me talking about you just now?"

"Don't remember. . . ."

"Do you remember seeing slides that you named for me?"

"Yes, I think so. . . ."

"When was that, Neil?"

"You brought that slide projector to my room last night, I remember now."

"What have we been doing this morning, Neil?"

"You gave me those shots with the hypodermic needle. . . . And then, I remember now, you showed me those slides again."

"Did you see anything unusual, Neil?"

"That's right, I saw that man walking his dog again. He got all of the way to the middle of the picture this time, I remember him getting all of the way to the middle."

"All right, Neil. You did have one of your little seizures. Everything is going fine, so just rest for a while."

Seizures such as Neil's often begin in one spot in the brain, where the damaged nerve cells are located. This site is called the focus. Then the seizures spread to surrounding, more normal brain. The first symptoms of the seizure, from the disturbed function of the brain in the focus, are often called the seizure "aura." Since seizures often start in the same spot, these aurae are frequently the same from one seizure to another. Not everyone with focal epilepsy has an aura. But many patients with foci in the temporal lobe have aurae such as Neil's: either a hallucination (a sensation of something that is not there) or an illusion (a distortion of something that is). Hallucinations may be visual, such as Neil's picture of a man and a dog at the start of each seizure. Another patient may see a bucolic farm scene at the start of each of his spells; other patients may hear voices, or music, or smell something (usually unpleasant). Very occasionally, an aura will consist of a specific event from the patient's past, a hallucination whose implications for our knowledge about human memory have been discussed. One common illusion at the beginning of a temporal-lobe seizure is fear, or feelings of unusual familiarity ("I've been here before," known as *déjà vu*). Objects may seem unusually far or near. One illusion that these patients are often reluctant to report involves alterations in the apparent size of things; they are afraid

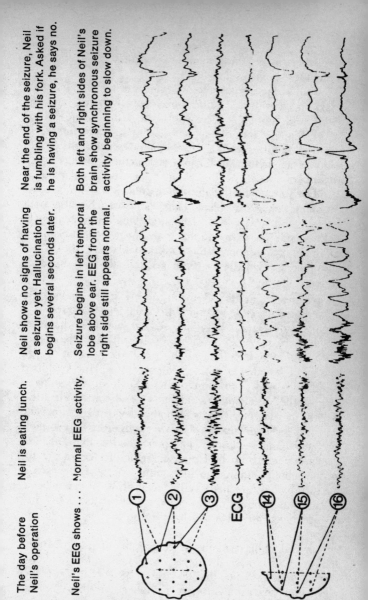

The day before Neil's operation

Neil is eating lunch.

Near the end of the seizure, Neil is fumbling with his fork. Asked if he is having a seizure, he says no.

Neil's EEG shows . . . Normal EEG activity.

Neil shows no signs of having a seizure yet. Hallucination begins several seconds later.

Both left and right sides of Neil's brain show synchronous seizure activity, beginning to slow down.

Seizure begins in left temporal lobe above ear. EEG from the right side still appears normal.

Brain waves before and during a psychomotor seizure.

that others might think them crazy if they admit that they sometimes see giants and little people.

Spread of the seizure into wider areas of the temporal lobe is associated with automatic movements: walking, fumbling with things, muttering a few words. When the seizure extends to the hippocampus, there is a loss of memory for recent events. But during such a "psychomotor" seizure, the patient seldom does anything elaborately purposeful or planned, such as cooking a meal, stealing a car, or (as in a popular novel) destroying a computer center. He is unlikely to harm another person, except by accident. (*The Terminal Man* was both epileptic and psychotic.)[1] "Temporary insanity" defenses are still sometimes used in criminal trials, the defendant claiming no memory for robbing the bank and blaming it all on a psychomotor seizure, even though there is almost no medical evidence that this ever happens.

Neil's seizure in the OR was one of his "little" seizures. He sometimes has a little seizure of his usual type, followed by a major motor seizure. This "big" seizure, which is what most people think seizures consist of, involves rhythmic jerking of both arms and legs; indeed the whole body seems to go through a sequence of stiffness followed by jerking, and then a relaxed period of unconsciousness. Such seizures seem to involve the simultaneous induction of abnormal electrical activity in widespread areas of both sides of the brain. The electrical "batteries" upon which nerve cells operate are severely discharged, and the unconscious period corresponds in part to the time needed for the battery-charging mechanisms (actually metabolic pumps, moving sodium and potassium across the membrane of the nerve cell) to reestablish normal voltages. Neil, reading about this earlier, observed that it sounded like a computer center "coming back up on line" after a power failure.

How do anticonvulsant drugs prevent seizures? Mostly, they seem to prevent the spread of seizures within the brain; there is little evidence that they get at the root of the matter and quiet down the primary pathology, in the way that levodopa is thought to work in Parkinson's disease. One anticonvulsant, sodium valproate, is thought to increase the stockpile of the neurotransmitter GABA in the brain. This neurotransmitter acts in an inhibitory manner, decreasing the electrical activity in downstream neurons. Thus the anticonvulsant might be acting by preventing the runaway conditions which occur when inhibitory neurotransmitter supplies become exhausted: a little seizure might still get started, but surrounding neurons would be more resistant to being fatigued into runaway activity.

The physiologic mechanisms that allow little seizures to get bigger by "taking over" normal neurons are reasonably well known, but why that spread occurs only occasionally, rather than all the time, is much less clear. Certain things seem to increase the likelihood of seizure spread in a particular patient; drowsiness and emotional stress are two fairly common ones. Some patients find little "tricks" that seem to prevent seizure spread. One patient found that thinking about something else the moment he detected the seizure would stop it. In another patient, touching an arm or the face may stop the spread of the seizure.

A more elaborate strategy to stop seizure spread is to let the patient learn to alter his own EEG by biofeedback. Usually this involves feeding back a signal to the patient that tells him how much of his EEG is in a particular frequency range. The patient then tries to find ways of changing that signal, usually shifting the EEG to higher frequencies. In an occasional patient, seizure rates decrease. But with seizures influenced by so many things, it requires a very carefully designed study to be sure that the treatment is really effective—whether it is actually the biofeedback, or the new drug, or the new operation that has decreased the seizures. Even bringing a patient into a hospital *and doing nothing else* often reduces seizure rate and severity, sometimes stopping seizures altogether for a while. This is the reason that controls are so essential to studies of treatments for seizures. In such rigorous studies, a number of old standbys and some widely publicized new treatments for epilepsy don't seem to do anything specific.

What would cause the rapid spread of Neil's seizure out of the left temporal lobe to involve many regions of both sides of the brain at once? The spread can be so sudden that the EEG experts often speak about "the wrong switch having been pushed," and there has been much speculation about the brain mechanisms involved. The seizure does not simply spread like a wave across the surface of the brain, one region involving its neighbor in the manner of a forest fire. It appears that the very mechanisms which the normal brain uses to control levels of arousal, to control sleep and wakefulness, have become involved in the seizure. These mechanisms are collectively called the reticular activating system and are known from several decades of animal experiments to reside deep in the core of the brain. They have widespread connections to all parts of the brain on both sides, like a manager with a finger in every pie.

The reticular activating system has two component parts. One part regulates overall arousal and the various aspects of the sleep-wakefulness cycle; another provides selective attention. The arousal part of the system, located in the upper part of the brain stem, acts like an amplifier placed in parallel with sensory inputs. When turned on, this amplifier takes those sensations and boosts them in order to "wake up" the cerebral cortex. A seizure spreading to this amplifier could rapidly spread to the whole cortex.

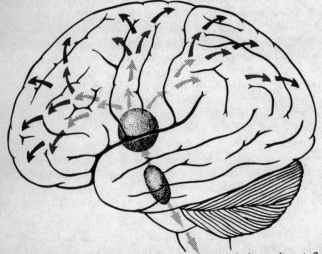

The reticular activating system: ascending and descending influences.

A small seizure from anywhere can spread into this amplifier. Should the seizure start in a piece of brain that has an easily detectable function, the first sign of the seizure will be a change in that function: a jerk of the hand if the seizure starts in motor strip, or a hallucination if it starts in temporal lobe; only after that warning will the rest of the brain become suddenly involved in the seizure. For many brain areas, however, there is no easily detectable function (no "local sign," as the neurologists call it) to serve as a warning.[2] The first indication of a seizure may only come after it has spread into the activating system: the stiffening and jerks of a full-blown seizure involving many regions of both sides of the brain. These "big seizures" are called by their French name, "grand mal."

Another common seizure also seems to involve the reticular activating sysem. In these seizures, the reticular activating system "amplifier" seems to be switched off briefly, rather than excessively activated as in big seizures. So the symptom of these seizures is a brief interruption of consciousness. This type of little seizure is called "petit mal." This seizure type typically comes on in early childhood, often changing its nature, indeed even sometimes disappearing as the child reaches puberty. Of all the seizure types, it is one of the few in which a family history of epilepsy is fairly common.

Patty's third-grade classwork had been getting worse. Her teacher and her parents considered her to be a normal eight-year-old, although given to "daydreaming" and "not paying attention." On spelling tests, there were gaps in the test paper, where she failed to write down two or three words. Patty said that she just didn't hear them. The words which she did write down were spelled correctly most of the time. Her physician was suspicious and sent her to a neurologist. As Patty talked, the neurologist noticed that she would stare fixedly for brief moments. An EEG later confirmed that these lapses were from brief seizures, involving both sides of the brain at the same time, causing brief periods where Patty was simply out of touch with the world around her, just as if a switch had been turned off and then back on again.

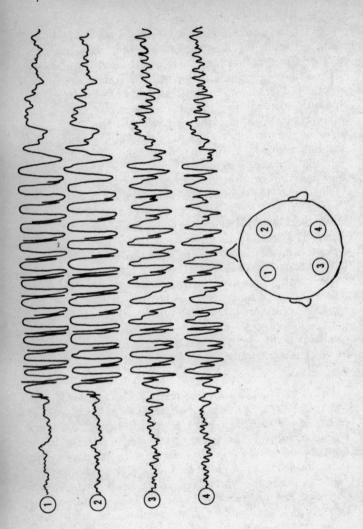

Patty's petit mal seizure lasts only a few seconds, during which she is out of touch with her surroundings. Note that the seizure starts and ends simultaneously on both sides of the brain, as one might expect if it were broadcast from the reticular activating system.

Patty was started on the anticonvulsant drug sodium valproate when it was released for use in the U.S. by the F.D.A., some years after its introduction in Europe. That drug has completely controlled her seizures and her schoolwork is finally back to normal.[3]

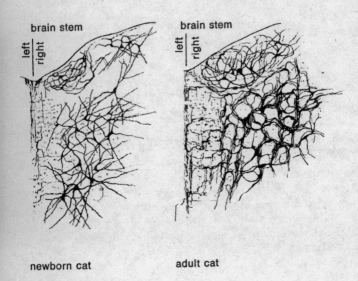

In the brain stem of the adult cat, dendrites have become rearranged into bundles which loop around longitudinally running groups of axons (modified from Scheibel and Scheibel 1973).

9

Paying Attention:
Inner Speech, Sleep, Schizophrenia, and Synapses

The reason that the activating system of the brain is so effective at spreading seizures around is its widespread connections to most parts of the brain. It has two components, arousal and selective attention; together, they are essential for higher brain functions.

Different sections of the arousal component of the reticular activating system regulate the complex processes of sleep. These sections are all located in the upper part of the brain stem. One section triggers deep sleep, also called "slow-wave sleep" because of the changes in the brain waves that go along with it. Normal awake resting brain waves have many oscillations at about eight to twelve cycles per second, called the "alpha rhythm" (a favorite of those who like to use biofeedback to condition brain waves). When the reticular activating system arouses a person, the alpha rhythm is replaced by activity that oscillates much faster and over a smaller voltage range. During deep sleep, the waves are much bigger and slower, oscillating one to two times a second, and leading to the name "slow-wave sleep."

This deep sleep is periodically interrupted by rapid-eye-movement (REM) sleep. These REM periods are triggered by another section of the reticular activating system. During REM, the brain waves oscillate more quickly, as though the person were awake; the eyes move about rapidly beneath the closed eyelids (hence the name), although other muscles are relaxed. If the person wakes during REM sleep, he often reports dreaming. REM sleep seems to be essential: if a subject is awakened every time he starts a period of REM sleep, there will be more REM sleep the next night. If such REM deprivation is kept up night after night, the subject's daytime performance will deteriorate much more than if the awakening had occurred in deep sleep. But why we need REM sleep, with its dreaming, is unknown.[1]

The arousal part of the activating system is located in the core of the upper portion of the brain stem. This site is rather vulnerable to damage from head injuries. This is probably because the brain stem acts much like the stem of a mushroom. During a blow to the head, the cerebral hemispheres may oscillate back and forth and put stress upon the brain stem, the way that a mushroom stem would suffer if the top of the mushroom were kicked. Brain-stem injuries often lead to the loss of consciousness because of trauma to the activating system; while the damage is often reversible, permanent coma sometimes results.

The second component of the activating system involves a more refined process: selective attention, rather than generalized arousal. This selective attention allows the discrimination of meaningful from unimportant stimuli: a mother responds to her baby's cry but not the noises of other children playing. It allows sustained concentration on one task, such as reading this chapter, with only a novel or imperative stimulus able to distract, such as a sonic boom or a call to dinner. Part of this component of the brain's activating system is located in the thalamus. It is the system altered by electrical stimulation during thalamic operations for Parkinson's disease, stimulation that shifts attention from the patient's internal world to what is coming in from the external world. This attentional shift enhances memory for what is coming in, while making it harder to call up information already in memory. The left thalamus in man has this property for verbal material, the right for visual-spatial information. In animals, this selective attention system shows little evidence of lateralized function.

It is tempting to speculate that this system should play a major role in conscious experience. It should determine to what we attend in the environment and the intensity with which it is remembered. When it switches away from external to internal stimuli, associations already in memory become available to compare with external stimuli, and as a source for internal modeling of different response strategies before taking action.

We often rehearse such internal strategies: how to respond to the boss when told of a promotion; what is the best way to get from the freeway to the restaurant for tonight's dinner engagement; how to design an experiment to determine the role of human alerting mechanisms. Each of these internally rehearsed models generates new associations from short- and long-term memory. This internal rehearsal often occurs while, elsewhere in the brain, external stimuli are receiving appropriate responses: the to-be-promoted-secretary types a letter, the driver continues down the freeway avoiding other cars, the experiment designer finishes showering.

The selective attention mechanism seems to be the system turning different brain areas on and off to the external or internal world. The internal-world aspect of conscious experience is enormously resistant to experimental study. Each of us knows that we have such an inner world, complete with "inner speech" and private ideas, but how does one establish the nature and existence of that inner world in someone else, or in another animal species?

One can imagine the diseases that would result if this selective attention system malfunctioned. With a minor disturbance of the system, the ability to hold attention on the external stimulus at hand would weaken, so that attention could be easily diverted by other external or internal stimuli. Such a defect is commonly seen in children with learning difficulties, children often labeled as "minimally brain damaged" or "minimally learning-disabled." Indeed, a proportion of these children show other neurologic signs suggesting that the thalamus or striatum may not be functioning normally.[2] Patients with hemorrhages in the left thalamus show a somewhat different defect. Attention can be sustained upon the external world, but things (words) from the internal world seem to intrude into external speech in an uncontrolled way, much to the patient's surprise and frustration, as in our patient who kept having the phrase "affirmative action" intrude on his attempts to name objects.

If the failure of the selective attention system were complete, one would expect a loss of all attention to external stimuli; the brain would be continuously turned in to its internal world. That seems to describe what often happens in autistic children, who show minimal or no attention to external stimuli. They act as though they are continuously attending to something inside themselves. But anatomical studies of the brains of autistic children show no anomalies, not even in striatum and thalamus, the areas where the selective attention system is thought to be located. Perhaps these studies have not looked for the right things, or perhaps our intuitive idea of what a damaged selective attention system would do is wrong.

A partial failure of the selective attention system might result in something in between minimal brain damage and autism. This seems similar to one of the defects described in schizophrenia. These patients "are brought to a topic totally irrelevant to the subject at hand, by any accidental thing that happens to affect their senses. . . . The normal directives through questions from without and purposive questions from within are incapable of holding the train of thought in the proper channels."[3] Schizophrenics may also blur the distinction between internal and external worlds. This is in contrast to the patient with a thalamic hemorrhage who seems to know this distinction clearly, even though the internal world intrudes on the external. Indeed, schizophrenia is a many-faceted disease, perhaps with multiple causes involving several different brain areas in different cases. Defects in selective attention are, at best, only part of the probelm.

A situation similar to acute schizophrenia can be produced in a normal person by an overdose of the drug amphetamine. Study of similar amphetamine overdosage in experimental animals has uncovered a clue to a neurotransmitter abnormality in the parts of the brain concerned with selective attention, a clue that eventually may be important in understanding what is wrong in schizophrenia. The culprit again seems to be the neurotransmitter dopamine—not too little dopamine, as was the case in Parkinson's disease, but rather too much effect from a normal amount.[4] A brief look at the connection between nerve cells, the synapse, and the ways it can be fooled illustrates how this seems to come about.

There are a number of steps in the process which one neuron uses to talk with another, a process called synaptic transmission. An electrical impulse in the first neuron travels down to the end of the axon, where there are regions specialized for releasing neurotransmitter molecules from the cell. A certain number of molecules are released by each impulse; the release process is triggered by calcium entering the cell following the voltage change of the impulse. The molecules start diffusing away from the cell membrane after their release, but the next neuron is just a short distance away, across what is called the synaptic cleft.

THE SYNAPSE

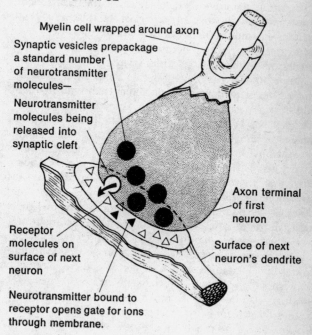

Myelin cell wrapped around axon

Synaptic vesicles prepackage a standard number of neurotransmitter molecules—

Neurotransmitter molecules being released into synaptic cleft

Axon terminal of first neuron

Receptor molecules on surface of next neuron

Surface of next neuron's dendrite

Neurotransmitter bound to receptor opens gate for ions through membrane.

The surface of the next neuron at the synapse is quite specialized for detecting the neurotransmitter molecules. There are a number of sites called receptors, where a chemical bond can be formed between the neurotransmitter molecule and the receptor molecule. The combination molecule then unlocks the membrane of the second neuron to allow certain substances through, such as sodium, potassium, or chloride ions. These ions, in turn, change the voltage of the second neuron. So the first cell releases "keys," the second cell has a number of "keyholes," and a key which fits the lock opens a "gate." There are also mechanisms in the synaptic cleft for cleaning up neurotransmitter molecules. Rather like a vacuum cleaner for keys, these "inactivation" mechanisms prevent some of the transmitter molecules from ever reaching the keyholes in the first place—indeed, they "suck the keys" back out of the "keyholes" after a short time, which results in the gates locking again to stop admitting chemicals to the second neuron. This is one way in which the effect of the impulse is limited in time: an impulse may last one millisecond itself, but its effect upon the second neuron via the neurotransmitter it releases may either be brief or long-lasting.

At one time, it was thought that in schizophrenia, neurons released too many dopamine molecules each time an impulse arrived at a synapse. However, the supporting evidence for that idea has been mixed.[5] Autopsy study of the brains of some schizophrenics suggests that there are many more receptors for dopamine in the striatum than in the same brain area of normal people.[6] There thus may be an excess of "keyholes," rather than more "keys." And since the usual situation is to have more neurotransmitter available than receptors (more keys than keyholes), the presence of more receptors means that each impulse has a greater effect on the next neuron.

In other diseases, too little neurotransmitter is produced—there are too few "keys," as in the shortage of dopamine in Parkinson's disease. In still other diseases, there may be too few keyholes. Apparently this happens in myasthenia gravis, in which the body's immune mechanism runs amok and destroys 80 percent of the receptor sites at the nerve-muscle synapse. There are problems when the cleanup process fails, so that neurotransmitters stay around too long. Some poisons work this way, such as nerve gases and the pesticides. Other poisons such as curare act as if they were keys which fit the lock but don't turn, their presence in the keyhole keeping the right keys from entering the lock.[7] Many drugs more gently affect the synapse by changing the number of keys, by partially occupying keyholes, or by slowing the cleanup.[8]

In one way or another, the effect of the impulse is being modulated to alter the strength of the functional connection between the two neurons. From such synaptic alterations come the hallucinations of LSD and the therapeutic actions of levodopa. What drugs modify the synapses in the memory or selective-attention pathways? No one knows.

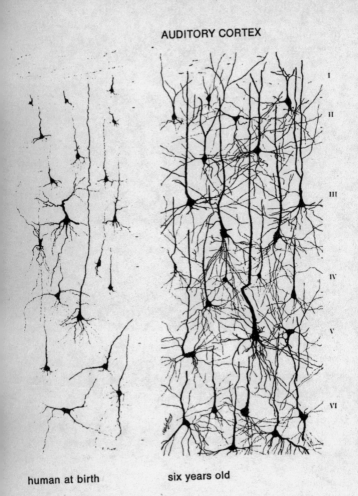

AUDITORY CORTEX

human at birth   six years old

10

The Brain Controlling Itself: Loudness Adjustments in Sensation

Switching different brain areas between internal and external stimuli is only part of the reticular activating system's role in selective attention. Another aspect is regulating the flow of information into the brain. This is accomplished through a series of "downstream" connections that act as though they were adjusting the sensitivity of sensory systems up or down, depending on the novelty or irrelevance of a particular sensation. This system is one of the mechanisms that prevents the brain from being continuously bombarded by the touch sensations produced by our clothes, but still able to detect the new touch of a mosquito. It was, in fact, one of the earliest functions of the reticular activating system to be discovered.

This system regulates pain sensation as well as touch, so that one way to control severe pain would be to change the reticular activating system so that it turns down the sensitivity in pain pathways. Some painkilling drugs may act this way, turning down the central "pain loudness" control.[1] Electrical stimulation of part of the reticular activating system is another way to turn down the intensity of pain input. Indeed, stimulation of the upper brain stem in animals seems to block pain completely. At least, the animals act as though they are not experiencing pain when their tails are shocked or overheated. Stimulation of this same area of brain stem has been used to treat a few difficult pain problems in man. In some of these patients, a few minutes of stimulation will lead to many hours of relief from chronic pain. In man, acute pain (as from a pinprick) is little altered by such brain-stem stimulation. Many treatments that work well for human acute pain, however, are of little value in chronic pain problems.

In both man and animals, the effects on pain from this electrical stimulation of the reticular activating system are blocked by the same drugs that block the function of opiates (morphine and its relatives). Stimulation locally releases substances with opiatelike properties. It has been suggested that these opiatelike substances are a class of neurotransmitters (endorphins, a word meant to suggest "internal morphine") which are used by the nerve cells in this system to control pain sensation. It has also been proposed that opiate drugs control pain through this same system.

Since the internal morphine pain control system can be stimulated electrically and by drugs, it can presumably be stimulated naturally by appropriate experiences. There are, for example, many war stories of soldiers who seem to feel no pain from a wound until after they have been evacuated from the battlefield. Many people who suffer from chronic pain disorders (as we discuss in Chapter 13) wish that they had a way of voluntarily activating this pain-control system. There is now evidence that the "power of suggestion" can act through this system. Physicians and faith healers alike have long known that for a certain percentage of pain sufferers, a confident attitude can alone work wonders ("Just take these pills and the pain will be gone tomorrow. You'll see!"). The "placebo" effect is very strong for chronic pain disorders; at least one-third of all patients will respond as well to a sugar pill as to narcotics, at least for a while.

When the evidence for an "internal morphine" system in the brain began to accumulate, several groups of researchers designed experiments around placebos, to see if the placebo activated the "internal morphine" and hence suppressed the pain.[2] One approach was to see if morphine-blocking drugs also blocked the placebo responses. In one experiment, dental patients were used; they were forewarned that "experimental drugs" were being used which might or might not help their pain, and they consented to the experiment on them. These dental patients had just had an impacted wisdom tooth extracted, a procedure which predictably causes postoperative pain for some time. After their local anesthetics had worn off and the period of postoperative pain had started, the patients were given two injections, some hours apart. Some of the patients ("placebo responders") got pain relief from the first injection, even though it was really nothing but water. The second injection was really a morphine antagonist called naloxone (typically used to treat heroin overdoses); it is the "wrong key," whose molecules plug up the "keyholes" on which opiate drugs (morphine, heroin, Methadone, Demerol) act. After this dose of naloxone, the placebo-responders again felt their pain. This result strongly suggests that the "power of suggestion" was acting through releasing an internal morphine.

A different type of pain problem looks as if it may be due to excessive amplification in pathways that convey sensation from transducers to the brain. What happens then is much like turning an audio amplifier gain up full-blast; the sensations that come through are distorted and unpleasant. Light touch feels like burning or an electric shock. This excessive amplification seems to occur when pathways for sensation from the body are damaged in the brain. Less sensation seems to reach the cortex, which may in turn ask the reticular activating system to turn up the amplifier full-blast. This is one possibility for what is wrong with the following patient.

It took about three months after her stroke for Edith to regain full function in her paralyzed left hand. Even when it moved well, the hand felt numb, as did the rest of her left side. Six months later, a pain problem had developed. The lightest touch caused her hand to burn and tingle. She couldn't even put the hand in her pocket, the touch of the fabric was so unpleasant. When the neurologist touched her left side with a pin, she felt it only when the pin was pushed hard, and then only after a delay. But when she did feel the sensation, it was awful; like a painful burning shock. The neurologist gave her a prescription that helped the pain somewhat, but then she developed a severe skin rash and had to stop taking the medicine. After that the pain was as bad as ever. No other medicines helped.

Three years later, an operation was done where an electrode was placed in her right brain in a part of the white matter which the surgeon called the "internal capsule," a main sensory pathway carrying body sensation to the cortex. When that electrode was stimulated, she felt a moderate tingling in the left

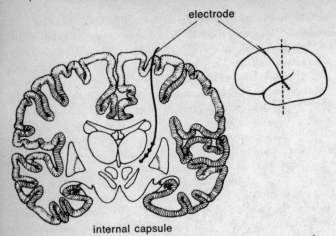

Edith's electrode in the internal capsule, stimulating axons on their way from the thalamus to the sensory strip.

side of her body. After a few days of nearly continuous stimulation, the pain faded away. The surgeon fixed the electrode so that she could turn it on and off through her skin, using a control like a miniature radio transmitter. She really didn't understand much about that, but she knew that if she occasionally used it during the day, her left hand would stop hurting, would not "burn" when touched, felt warm again, and that she could use her hand for most things.

Edith developed what is known as the "thalamic" syndrome after her stroke, so named because a stroke damaging the sensory pathways in the thalamus is one cause of this pain condition (it was once thought to be the only one; now we know that damage to sensory pathways almost anywhere in the brain can produce the same pain symptoms). With damage to the sensory pathways, the amplifier seems to be turned up so that when sensation does get through, it is too intense: burning, shocking. One way of treating this particular kind of pain, turning the amplifier setting of the sensory system down toward normal, is to increase the other sensations reaching the cortex. It is thought that the stimulating electrode in the internal capsule does this by artificially activating the axons that normally carry sensation through the internal capsule to the sensory cortex. This, then, could change the setting of the amplifier part of the reticular activating system indirectly, by altering the reports given it by other brain areas on the level of sensation they are receiving. Modern technology allows such an electrode to be activated by a radio receiver implanted beneath the skin; the receiver in turn is activated by a small radio transmitter whose antenna is placed on the skin over the receiver.

Problems like Edith's with the "volume control" in sensory systems represent only a very small percentage of the cases of chronic pain. Manipulation of this amplifier-setting system by stimulation is a method of treatment still undergoing evaluation; it seems to be of value for only a few pain problems. We will return later to more common causes of persisting pain in which the amplifiers are set normally, such as those in which seemingly normal pain signals are continuously generated by damage to peripheral nerves.

The ability to discriminate novel from repetitious sensations is so important that the amplifier controls in the reticular activating system are not the only mechanism in the nervous system entrusted with this job. Mechanisms which minimize repetitive sensations exist at many levels, beginning with mechanisms built into single nerve cells. As someone once noted: "Big brains, like big governments, do not do simple things in simple ways."

How do individual neurons detect the novel and ignore the mundane? An external stimulus starts up neural signals in a transducer neuron by producing a voltage change proportional to the stimulus strength. This transducer voltage may sag with time, becoming only half as big after a second or two, even though the stimulus strength is being held constant. In some transducers, this sag is dramatic: bending a hair on your skin produces a transducer voltage for only a few seconds. In other transducers, such as those in muscles and joints, the sag (or adaptation, as it is formally called) is less than 50 percent and also takes longer to occur.

A second opportunity for reducing the intensity of the neural message sent occurs when the transducer voltage produces impulses at the beginning of the transducer neuron's axon. The rate at which impulses are produced is often proportional to the transducer voltage, although there is a minimum level (or threshold) for impulse production. This impulse production rate may also sag, even if the transducer voltage is constant. Thus there are two opportunities for reducing the message sent (the impulse train), even before the message reaches the spinal cord or brain. Changes in such mechanisms occur in inflammation, when transducer neurons in the skin adjacent to an injury become more sensitive, producing impulse trains more easily and for longer times than normal.

When each impulse reaches the spinal cord (or an analogous part of the brain stem), it releases a packet of neurotransmitter molecules onto another neuron, the second neuron in the sensory chain. During a long train of impulses, the amount of neurotransmitter released per impulse may change. In some cases, it goes up (facilitation or potentiation); in other cases, it goes down (antifacilitation, sometimes called synaptic depression).[3] In some cases, the amount of neurotransmitter released per impulse is simply under the control of the historical factors within the axon terminal itself, such as fatigue. In other cases, there is regulation of the synaptic transmission between two neurons by a third neuron.

An excellent example of the regulatory processes has been studied in a group of neurons in the sea slug *Aplysia*.[4] This animal has a simple withdrawal reflex, quite analogous to the one humans use to lift a foot after encountering a thumbtack. The gill will be withdrawn if some adjacent skin is touched. If the skin is repeatedly touched every twenty seconds, the animal will cease withdrawing its gill so vigorously. This decline in the response to a repeated stimulus is called "habituation" when it can be reversed by presenting a novel stimulus, such as touching the animal's head. Following that novel stimulus elsewhere, the gill-withdrawal reflex will be restored to its original vigor.

The decline in the effectiveness of the skin stimulus occurs because, after a number of stimuli, the transducer neuron's axon terminals release less and less neurotransmitter per impulse. The novel stimulus to the head does not stimulate the

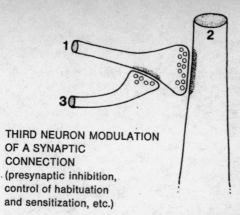

**THIRD NEURON MODULATION
OF A SYNAPTIC
CONNECTION**
(presynaptic inhibition,
control of habituation
and sensitization, etc.)

skin transducer neurons involved in the gill-withdrawal reflex; how then is the reset message delivered to their axon terminals, that something has touched the head? The answer is thought to involve an interneuron, which receives the information from the head but which regulates the synaptic transmission from the transducer neuron to the second neuron of the reflex chain (which is the motor neuron activating the gill muscles). This interneuron affects the axon terminal of the transducer neuron, changing the amount of calcium which enters the terminal upon the next arrival of an impulse from the transducer. More calcium causes more transmitter to be released; since calcium entry declines during habituation, this reverses the habituation.

Another use of an interneuron is involved in "presetting" an axon to release less neurotransmitter when an impulse arrives; this interneuron seems to make a connection to the axon terminals of the transducer neuron and has the action of reducing the neurotransmitter released by an impulse when one arrives. This is called "presynaptic inhibition." Its effect is to reduce the transmitter released by an impulse by a certain percentage, so it is analogous to division (in contrast to postsynaptic inhibition, which is analogous to subtraction).

The anatomical and physiological aspects of presynaptic inhibition have been studied in many types of animals. This

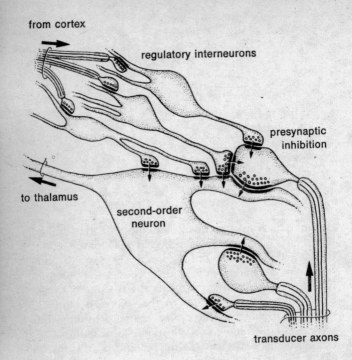

Interneuron regulation of sensory messages. Transducer neurons in the hand send their axons to second-order neurons which are located at the base of the brain, near its junction with the spinal cord. Their axon terminals make synapses onto a second-order neuron whose axon, in turn, ascends toward the thalamus. This synaptic transmission is regulated by other neurons ("interneurons") whose axons end upon the transducer axon's terminals to produce presynaptic inhibition. The regulating interneurons are activated by inputs descending from the cerebral cortex (such as from the sensory and motor strips). Other interneurons produce postsynaptic inhibition and excitation directly upon the second-order neuron. Commands descending from cortex determine the intensity with which sensory messages are relayed onward to the thalamus and cortex; they also regulate how large an area of skin will activate a given second-order neuron (diagram modified from Walberg, 1965).

interneuron regulation of impulse-evoked neurotransmitter release seems to be a widespread principle of the organization of nervous systems. Whether or not habituation, or perhaps learning, is also mediated through such presynaptic regulation is not yet known in many cases. Like the simpler adaptation, habituation may occur at many stages of the neuron chain, perhaps using multiple mechanisms.

Thus, not only do "loudness controls" exist within single neurons, but there are also systems of neurons which may regulate the internal settings. A similar scheme has been hypothesized for learning: if synapses (or other aspects of neurons) are to modify themselves with use, surely there will exist systems of neurons which will regulate this plasticity, which will permit or prohibit plasticity under varying circumstances.[5] For example, one role of the hippocampus in short-term memory might be to regulate the facilitation properties of synapses elsewhere in the brain; with destruction of the hippocampus, the synapses elsewhere might fail to receive the necessary permission to be modified by use, and thus a new memory trace could not be established.

VISUAL CORTEX

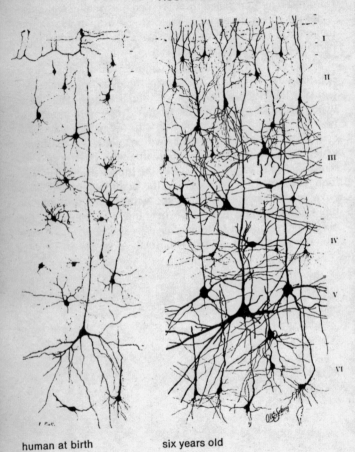

human at birth   six years old

11

Circuits for Seeing:
Nature vs. Nurture in
Brain Development

The screen goes blank after Neil finishes with the last slide. Immediately, everyone in the OR moves around a little and the noise level rises, as all had been quiet during the performance and recording.

"Very nice, Neil. Want to move around a little?" asks the neurosurgeon.

"Aagh! My leg fell asleep. Now my toes are tingling. Is that important? I wouldn't mention it, but you told me to tell you if I felt something," says Neil.

"Sounds familiar. That happens to nerves when they get blocked by pressure. No problem as long as they're tingling and not numb. The anesthesiologist will check out things. Just move around a little."

There is a chain of neurons which links the skin with the sensory strip of Neil's cerebral cortex. First is the transducer neuron in the skin, which senses pressure, the movement of a hair, or temperature. Next comes the "second-order" neuron, across the synapse in the spinal cord or in the brain stem. The third neuron in the chain is often in the thalamus. The axons from the thalamus go up to the sensory strip of cerebral cortex, so that there are at least three neurons in the chain before a cortical cell gets a chance. The direct four-neuron chain is not, however, the only path. As we have seen, there may be a relay in the brain stem's reticular activating system. Some axon branches go up to the cerebellum, the elaborate infolded structure which sits atop the brain stem and assists in the coordination of movements and posture. It too talks to the sensory strip. Each of the senses has a similar set of direct and indirect pathways from the transducers to the cerebral cortex.[1]

It is customary to think of this as a relay race, with something

113

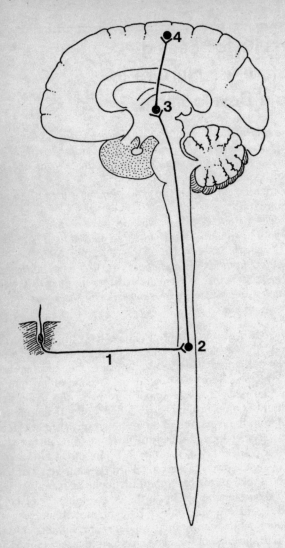

Chain of neurons conveying hair-movement sensation to spinal cord (or brain stem), thalamus, and cortex. Each sensory system has a similar chain, sometimes involving more neurons.

simply being passed on from one neuron to another. It is, however, more analogous to the spread of a rumor. As each successive neuron retells the story, it may leave out certain details, add others, or perhaps not retell the story at all. Each neuron has its own characteristic style of storytelling. Particularly in the case of vision, neurophysiologists have been able to partially understand what happens at each successive neuron in the chain. The transducer neurons are the rods and cones

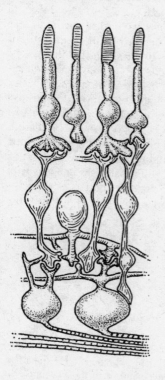

Transducer neuron
(rod or cone) converts
light intensity
into voltage.

Second neuron is
the bipolar cell.

Third neuron is the
retinal ganglion cell.

Axons go in optic
nerve to thalamus.

The first few stages of the neuron chain in the visual pathway are located in the retina at the back of the eye. Contrast between adjacent regions of the visual world is enhanced by the actions of additional cell types, such as the octopus-shaped cell in the center of the drawing (drawing modified from Dowling and Boycott 1966).

of the retina. The second and third neurons of the chain are also in the retina. The axon of the third neuron, called the retinal ganglion cell, runs all the way from the eye to the thalamus. The axon of the fourth neuron runs from the thalamus up to the visual cortex in the back of the brain.

The first neuron, the rod or cone, is very small: it responds only to light shining directly upon it. Diffuse light is one easy way to stimulate a single rod or cone; if a small spot of light is used (such as when looking at a star), it must be centered directly upon the particular transducer cell. So the image of the visual world projected upside-down on the back of the eye is changed into the voltages of a mosaic of transducer neurons.

The second and third neurons in the chain start the process of comparing neighboring regions.[2] A particular neuron responds best to a small spot of light, covering a number of adjacent rods and cones. If the light spot moves, the positive voltage changes into a negative one. If one moves the spot even farther away, the voltage becomes very small. So there is a doughnut-shaped map: the transducers in the center cause excitation, those in the surrounding ring cause inhibition.[3] What happens if both areas are illuminated, as happens with diffuse light? The positive and negative influences cancel—sometimes totally, as if the light were not there at all, sometimes only partially.

The RIGHT visual cortex gets messages from both LEFT and RIGHT eyes about images from the LEFT visual field (everything to the left of the fixation point). All axons from the retina (third-order neurons) representing LEFT visual field go to the RIGHT thalamus. In connecting with the fourth-order neurons in the thalamus, the messages from the two eyes are kept separate and not mixed. This again happens in layer IVc of visual cortex; the axons of the fifth-order neurons, however, converge upon sixth-order neurons nearby, which then have the opportunity to compare the somewhat different images seen by the two eyes.

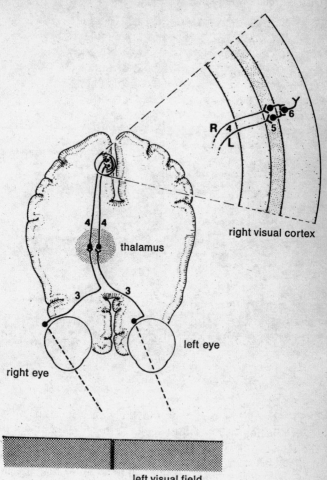

right visual cortex

thalamus

right eye

left eye

left visual field

TESTING A RETINAL NEURON
center-surround type

no response from neuron
"Nothing there, let's try moving the spot of light around."

excellent response
"Ah, there's the right spot for this neuron. We'll mark the spot with a cross."

poor response
"Making the spot bigger seems to inhibit this neuron."

very good response
"A bar of light works just about as well as a spot."

very good response
"This neuron isn't sensitive to the orientation of the line, either."

TESTING A CORTICAL NEURON
simple type

no response

good response

excellent response
"This neuron seems to like a long white line."

no response
"But it's very particular about the orientation of the line."

no response
"Even at the optimal angle of orientation, it's particular about the width of the white line."

The fourth neuron,[4] in the thalamus, is even more discriminating against diffuse light. Essentially, the image of the visual world has been changed by emphasizing the borderlines between darker and lighter areas. Such a neuron may, however, respond equally well to a spot or line: it may not be able to tell the difference between such symbols as ·, 1, /, −, +, etc. This is also true for the fifth neuron, located in the visual cortex. Starting with the sixth neuron, a voltage change can be evoked only if the orientation of a borderline is just right. If the animal is looking at a narrow line, such as the letter "l," a train of impulses may be evoked in some sixth-order cells. The line must be located in a certain part of the visual field, depending upon where the cell is located in the visual cortex, as there is a map of the visual world spread out across the visual cortex. Besides being in the correct place in space, the line must also be tilted at the correct angle. Orientation matters a lot to a sixth-order nerve cell (but any orientation will do in fifth-order cells and those earlier in the visual pathway). Some sixth-order neurons prefer vertical lines, others horizontal, others some particular angle. If the line is rotated 10° from the preferred orientation, the impulses are likely to cease in that cell, but start up in another group of neurons.[5]

The seventh-order cells often have even more complex properties. They, too, are particular about the orientation of a line but they are not as particular about its position in space. Some, called hypercomplex neurons,[6] are particular about the line's length, however. Thus, some are specialists in short, horizontal lines. They could tell the difference between — and ——.

Thus the neurons of the visual system are not equally interested in everything one sees. They take apart the cameralike visual image presented to the mosaic of retinal transducer cells. It turns out that they take it apart in ways that extract information about borderlines between lighter and darker areas. If one looks at a checkerboard, most of the response of the visual brain is devoted to the lines forming the edges of the black and red squares. Many fewer cells seem to be needed to process the information about the centers of the squares. The visual cortex is wired up from birth to accomplish all of this; however, the wiring can be modified by the experiences of infancy, as the following case illustrates.

Ross is six years old and cross-eyed. He seems almost blind in his right eye. But it now is too late to do much about it. Cross-eyed babies who see doctors regularly will be sent to an eye surgeon, as most doctors know that delays in correcting the defect will mean that the child may become functionally blind in one eye or the other. Lack of money in his family is not really the reason that Ross has now lost the sight of his right eye. *Ignorance is expensive. His mother simply thought of correcting crossed eyes in the same cosmetic terms as fixing a crooked nose.*

**TESTING A
CORTICAL NEURON
complex type**

**TESTING A
CORTICAL NEURON
hypercomplex type**

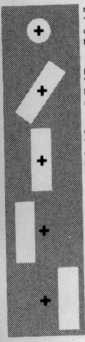

no response
"This neuron
doesn't seem
to like spots."

poor response
"A line isn't much
better. Let's try
rotating the line."

excellent response
"There's the
optimal orientation
for the line."

very good
response
"This neuron isn't
very particular
about the exact
position of the
line."

very good
response
"So this neuron
specializes in
vertical lines."

no response

good response
"That's the optimal
angle of orientation."

excellent response
"But short lines
stimulate this
neuron better
than long ones."

no response
"Even with the
short line,
it's still got to be at
the right orientation."

very good
response
"Looks as if short
45° white line
works well in a
number of different
positions."

That one shouldn't wait around too long with a cross-eyed child was just one of those empirical facts in medicine until a few years ago. Then, basic science finally advanced to a state in which techniques and knowledge were available to address the question: how does experience modify the brain's wiring? One of the first disorders which was illuminated by the research results was the mechanism by which the brain comes to ignore an otherwise good eye.

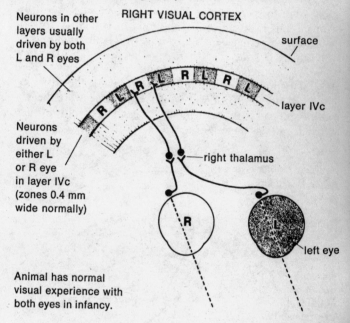

RIGHT VISUAL CORTEX

Neurons in other layers usually driven by both L and R eyes

surface

layer IVc

Neurons driven by either L or R eye in layer IVc (zones 0.4 mm wide normally)

right thalamus

R

L

left eye

Animal has normal visual experience with both eyes in infancy.

The two eyes see slightly different views of the world, as the reader can appreciate by alternately closing one eye and then the other, while reading this page. The "rangefinder" effect is quite useful for judging distances, at least out to about five meters away (the length of an automobile).[7] Not until the sixth-order nerve cell is the information from the left and right eyes combined, at least in primate visual cortex. At lower levels of the visual system, a cell either responds to light shown in the left eye, or to light shown in the right eye.

In the cortex, there are cells responsive to both eyes. The visual cortex is wired up from birth with this ability, but it can be lost if the infant does not get a chance to use both eyes *together*.

An infant monkey with a frosted contact lens, switched daily from one eye to the other, gets visual experience with each eye but the two eyes never get a chance to work together. The cells which would ordinarily respond to both eyes now seem to specialize in one eye or the other. Even if the contact lens is discarded and the monkey is then given normal visual experience for a long time with both eyes together, the situation tends not to reverse: there will be very few coordinating nerve cells, permanently. There seems to be a critical period in infancy (in monkeys, about the first nine months after birth) during which experience with the environment can make permanent wiring changes, at least such obvious ones as these.[8]

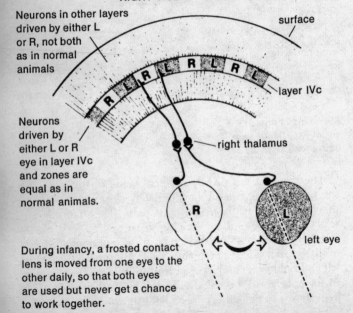

RIGHT VISUAL CORTEX

Neurons in other layers driven by either L or R, not both as in normal animals

surface

layer IVc

Neurons driven by either L or R eye in layer IVc and zones are equal as in normal animals.

right thalamus

left eye

During infancy, a frosted contact lens is moved from one eye to the other daily, so that both eyes are used but never get a chance to work together.

Such research, of course, has many implications for our own infants. Babies born cross-eyed have a problem similar to that of the infant monkey just noted; the two eyes never get a chance to work together. Rather than seeing double, humans with a squint will tend to suppress permanently the vision in one eye, becoming functionally blind in that eye (monkeys, interestingly, do not have this problem if made artificially cross-eyed; they can switch back and forth from one eye to the other). Such animal research has emphasized the need to correct cross-eyed conditions in infants as soon as practical; the relevant critical period in humans is probably the first two years after birth.[9] There are also indications that there are critical periods for other brain systems besides the visual system. The timing of each system's critical periods is often different. Indeed, there is a hint that such critical periods may exist for human language.[10] The existence of critical periods for normal brain development blurs the distinction between environment and heredity: now, early environment as well as heredity determines the brain machinery that is available for later interactions with the environment.

As David Hubel noted, "Perhaps the most exciting possibility for the future is the extension of this type of work to other systems besides sensory. Experimental psychologists and psychiatrists both emphasize the importance of early experience on subsequent behavior patterns—could it be that deprivation of social contacts or the existence of other abnormal emotional situations early in life may lead to a deterioration or distortion of connections in some yet unexplored parts of the brain?"[11]

The current thinking among researchers is that during the critical period, there is competition among nerve cells in the visual cortex, a variation upon the survival of the fittest. If the frosted contact lens is left on one eye for a while during the first six months of life, the usual 50-50 representation of left and right eyes will change in the narrow band in layer IVc of the visual cortex where the fifth-order cells are located. Fifth-order cells normally specialize in either left eye or right eye, never both. Normally, there is a territory about 0.4 mm wide containing only cells representing the left eye, then a 0.4 mm territory for the right eye, alternating left-right-left-right every 0.4 mm across the visual cortex. In the infant monkey with

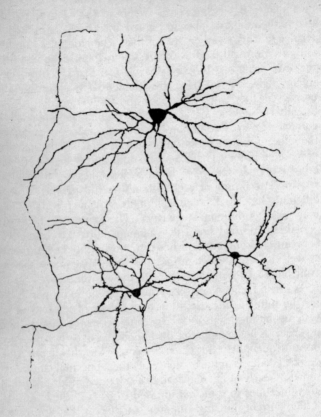

Some of the small neurons of the visual cortex in the monkey. The two neurons at the bottom are in layer IVc, where the inputs from the two eyes alternate left-right-left-right every 0.4 mm and are probably fifth-order neurons with circular receptive fields of the center-surround variety. Each neuron specializes in one eye or the other, depending on location. The neuron at top is in the layer immediately above; such cells receive inputs from both eyes, probably via axons (such as the one shown at left) from the small neurons below. They have "simple" line-orientation type receptive fields. (from J. S. Lund 1973).

badly blurred vision in one eye from the frosted contact lens, the territories from the "bad" eye shrink and those for the good eye expand as if the active "exercised" cells had won a competition for space in the brain.

There is even a problem with animals growing up in vertically striped environments: some researchers report that the orientation preference of the cells may shift, producing a higher than normal proportion of cells specializing in vertical orientations. As with other effects summarized above, there is now a new round of scientific literature, pointing out the exceptions and variations and occasionally reporting failures to reproduce other researchers' results. (One famous failure to reproduce others' findings from cat visual cortex turned out to be due to slightly different breeds of Siamese cats in Massachusetts and Wisconsin; that discovery in turn has opened up new opportunities to study the wiring accidents during prenatal development from which these cats suffer). Thus a picture emerges of how the cells are wired together, how their internal electrical processing of information is carried out, where they send their new messages, and what happens next.[12]

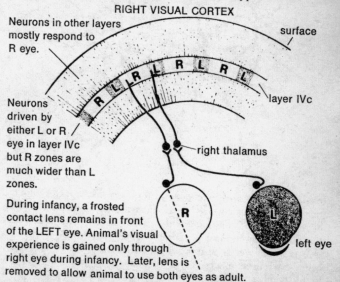

RIGHT VISUAL CORTEX

Neurons in other layers mostly respond to R eye.

surface

layer IVc

Neurons driven by either L or R eye in layer IVc but R zones are much wider than L zones.

right thalamus

During infancy, a frosted contact lens remains in front of the LEFT eye. Animal's visual experience is gained only through right eye during infancy. Later, lens is removed to allow animal to use both eyes as adult.

R

left eye

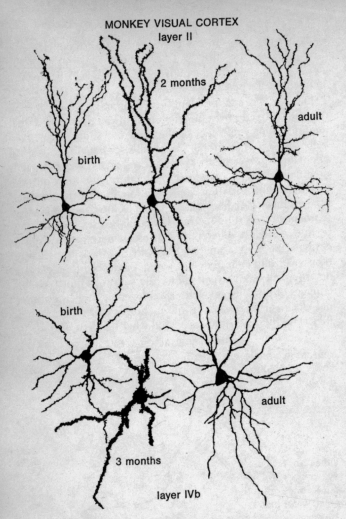

MONKEY VISUAL CORTEX
layer II

2 months

birth

adult

birth

adult

3 months

layer IVb

Dendritic spines greatly increase in numbers (budding?) during the first few months of postnatal development. Their numbers then drop by half in the following months. This suggests one anatomical substrate for the critical periods for visual experience. Upper cell type is from layer II of monkey visual cortex; the layer IVb cells below have been rotated 90° to fit onto the page (modified from Lund, Boothe, and Lund 1977).

12

Inside the Neuron: Making Decisions, Sending Messages

As can be seen from the hierarchy of neurons in the visual system, one function of neurons is to synthesize a new view of things, to take all the thousands of inputs to the neuron and combine them in a way that tells a new story, performs a new function, plays off one influence against another. The other task of the neuron is to move the message over some distance to the next neuron in the chain. Writing a letter does little good without a postal system to deliver it. Sometimes the next neuron is just next door, sometimes there is a long distance which must be traveled (from the tip of the toe, a single neuron runs all of the way to the base of the brain).

Whenever the message must travel more than about 1 mm, neurons use a special long-distance mechanism. The impulse, a rapid change in the voltage across the membrane of the neuron, starts the process. Every 1 mm (a half-space on most typewriters) along the neuron's axon there is a booster station which reproduces the impulse. It is as if a message were to be sent from one city to another by means of stationing soldiers on each hilltop along the way. The order is given and the first soldier fires his gun. When the next soldier hears it, he fires his gun, and so on until a gunshot is heard in the distant city. This is very much the scheme used by a neuron to get a message from one end of its axon to the other: the first impulse sets off an impulse 1 mm down the axon with about a 20-microsecond delay, that impulse sets off an impulse at the next relay station another 20 microseconds later, and so on. To cover a distance of 1 meter, it will take a total of 20 milliseconds (1/50 second, a typical shutter setting for a camera). The speed with which the message travels is 50

meters per second (180 kph, which is about 110 mph). The distance between booster stations varies, as does the delay before the booster station responds with its own impulse, so speeds vary from 1 to 150 meters per second. Some extremely thin axons are even slower, which is why the pain is delayed several seconds after you stub your toe.

This scheme would seem to have some weaknesses. In the analogy, suppose one soldier in the chain fell asleep, or just missed hearing the shot? What if the soldier mistook another loud noise for the signaling shot, and fired? Nerves also fail to conduct messages, and occasionally they generate false alarms.

Betty is twenty-nine and her work as a computer programmer has been disrupted lately by some sharp pains in her face. Sometimes the slightest touch to her lips will cause an agonizing, knifelike pain shooting down her jaw. Her neurologist knows that the symptoms match those of a disease called tic douloureux. But the neurologist is suspicious, because tic douloureux is a disease usually seen only in older people, and it is uncommon in someone who is only twenty-nine. A careful neurological examination is done, flashing lights here and there to see if there are any abnormal blind spots in her visual fields. Sure enough, there are two small areas where she cannot see lights. The diagnosis is multiple sclerosis, a disease of young adults affecting the insulating material around nerve cells. The neurologist begins treating her with one of the drugs which is sometimes useful in stopping such pains.

Betty's case illustrates two aspects of malfunctioning in the booster-station system used for nerve impulses. The simplest malfunction occurs when the message is simply blocked; the neurons in her eye are producing impulses in response to the light flashes, but the impulses are not being repeated in one short region of an axon where the insulating cells are missing, and thus no message arrives in the brain. The second malfunction is the one seen in Betty's face: an innocent touch is producing pain. The production of impulses without a stimulus, and of the overproduction of impulses upon a mild stimulus, is analogous to the soldier firing in response to noises in the night and raising a false alarm.[1]

Blocking the message can be done in a number of ways: arms "fall asleep," dentists use local anesthetics. In order to understand how the blockage occurs, one must move away from soldier-relay analogies and look at the way in which impulses are produced by the sodium and potassium ions in the saltwater solution which bathes nerve cells.

Atoms of sodium and potassium tend to lose one of their electrons, disrupting their neutral balance of positive and negative internal charges so that they become, on balance, positively charged "ions" (written Na^+ and K^+). Chloride tends to pick up a stray electron, thus it becomes a negatively charged ion (Cl^-). Most of the salt in a salt solution is really such a collection of positively and negatively charged ions. If they intermingle, as is the usual case, overall electrical neutrality of the solution is maintained. If something segregates them, such as the membrane of a neuron, then the separation of positively and negatively charged ions creates a voltage. It takes some work to segregate the ions, which is one of the things that the food we eat is used for (all cells of the body use their membranes to perform this segregation, not just nerve cells). Once segregated, the attraction of positive charges for negative charges tends to make the ions move, trying to intermingle again whenever the membrane of the neuron will allow it. There is ten times the concentration of sodium outside the cell as inside, thanks to pumps which keep throwing sodium ions out of the cell. The pumps are slow, rather analogous to charging a battery.

outside

membrane

inside

Na+
pumped out

K+
pumped in

The sodium concentration (Na$^+$, stars) inside the neuron is kept low by a pumping mechanism in the cell membrane; it captures sodium ions inside the cell and evicts them. At the same time, any potassium ions (K$^+$, squares) which can be captured outside the cell are carried inside, thus maintaining the high internal concentration of potassium. Tenfold and greater differences in internal and external concentrations are maintained by such pumps, which are driven by metabolic energy. Such concentration differences constitute the "batteries"; impulses and synaptic activity can selectively allow puffs of sodium to enter the cell and raise the internal voltage. Alternately, potassium can be released in small amounts to lower the internal voltage (see next figure).

Indeed, the membrane of a neuron is in the business of regulating ion flows. If it allows sodium ions to diffuse through the membrane, the net flow will be inward. If the membrane allows potassium ions through, the net flow will be outward. The membrane permeability can change rapidly, and movement of a small number of ions can produce large shifts in voltage.

An impulse is caused by a small patch of membrane becoming highly permeable to sodium ions. Thus, a puff of sodium ions enters the cell. Since each is positively charged, the inside of the cell becomes positive. The voltage reaches about 100 millivolts (1/10 volt) locally. This increase is the signal for the membrane to switch off its sodium permeability and to start letting more potassium ions through. Now a puff of potassium ions leaves the cell. This means positive charge is leaving the cell—different positive charges than those that entered the cell as sodium ions, but nonetheless positive charges. Therefore, the 100 millivolts falls back toward zero. The potassium channel tends not to close until the voltage reaches zero, so that the resting condition is reestablished. This swing in the membrane voltage from 0 to 100 and back to 0 takes about 1 millisecond (1/1000 second, about the fastest setting available on even expensive cameras). This voltage-time sequence is what is called the impulse. [2]

But what starts it? What causes the sodium channels to open up in the first place? It turns out that channels are sensitive to the voltage across the membrane. As the voltage increases a little, both sodium and potassium channels open up a bit more, but the resulting flows (sodium in, potassium out) largely cancel each other out. That is, until a certain voltage is reached, where upon the sodium flow becomes larger than the potassium flow. This particular voltage is called the "threshold" for the impulse. Beyond this level, there is a runaway condition, with sodium ions bringing positive charges into the cell much faster than potassium can carry positive charges out of the cell. This is the way that the impulse rises. It is analogous to pulling the trigger on a gun: for easy pulls, nothing happens; then a threshold is reached and off it goes.

But what causes the voltage to rise originally, so that it

Puff of sodium ions (Na+, stars) enters the neuron through an open sodium channel. This makes the cell interior 100 millivolts positive at the peak of the impulse.

To terminate the impulse, the sodium channels close. The potassium channels open and allow a puff of positive charges (K+, squares) to leave the cell. The internal voltage thus falls back to zero to terminate the impulse.

crosses this threshold voltage and sets off this sequence of events called the impulse? If we are talking about one of the booster-station sites in the middle of the axon, the voltage is simply the result of having another impulse occur 1 mm away. If one booster station (they are really called "nodes of Ranvier") has shifted its voltage by 100 millivolts, it will attract negative charges such as chloride ions away from the interior of the adjacent node. The loss of a negative charge has the same effect as the addition of a positive charge: it raises the voltage. So the neighboring node will experience a voltage shift too—not of 100 millivolts, but of some lesser amount. When the threshold voltage is reached at the neighboring node, it too experiences an inward puff of sodium ions which raises its voltage to 100 millivolts, and it too releases a puff of potassium ions to restore itself back to 0 after a millisecond or so. Just like the soldiers pulling their triggers on successive hilltops, this process repeats itself over and over at each successive node until the end of the axon is reached.

But what causes the very first impulse to occur at the very first node? Where does *that* voltage shift come from? What precedes the impulse? If the neuron is a transducer, the shift comes from the sodium channels in the membrane being opened by light, or mechanical pressure, or chemicals (depending on whether the transducer neuron is sensitive to light, touch, or taste). For more ordinary neurons inside the brain, the shift comes from channels in the membrane which have been opened by the arrival of a neurotransmitter chemical from an upstream neuron. As we have seen, an impulse arriving at the end of an axon releases such neurotransmitter chemicals. They diffuse the short distance between neurons, combine with receptor molecules, and open up the ion channels in the downstream cell. Some neurotransmitters open predominantly sodium channels, causing the voltage to swing positive; others open potassium (or chloride) channels, causing the inside of the cell to become more negative. Most neurons receive a mixture of such influences. Some synaptic inputs tend to produce positive voltages (called excitation), while others create negative influences (called inhibition). The balance of these voltages is what counts. If the balance is large enough to cross the threshold, an impulse will start.

Each neuron adds together thousands of such influences, but not all have an equal vote; some inputs are more powerful than others.

The balance not only determines if an impulse will be initiated, but also sets the production rate. This is analogous to pressing upon a sewing machine's control pedal: the harder

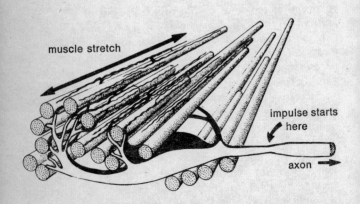

STRETCH TRANSDUCER NEURON converts elongation into a voltage change. The neuron's dendrites intertwine with the muscle fibers; when they are stretched, the dendrite's sodium channels open and positive charges enter the dendrite and raise its voltage. Impulses do not start in the dendrites, as the impulse threshold is very high there. The dendritic voltage, however, attracts charges away from the initial segment of the axon, thus raising its voltage somewhat. When its threshold is exceeded, impulses start there and spread from this "trigger zone" down the axon. The impulse production rate increases when the dendritic voltage is raised by more stretch. Such transducer neurons send their axons to "motor neurons" which increase the tension of other muscle fibers running in parallel; this system helps to maintain posture automatically, as a stretch by an external force leads to contraction to maintain the muscle's length. Such a "stretch reflex" is what physicians test with their rubber hammers. The transducer neuron in the diagram is from a crayfish (diagram modified from Florey and Florey 1955).

you push, the faster it stitches. Some neurons are capable of producing 1,300 impulses per second, although a more typical maximum rate is about 300 impulses per second. The rate tells the next neuron down the line what the balance was.

Another way of sending a message, besides using average impulse-production rate, is to use impulse patterning. Clustering the impulses into a burst, instead of spreading them out evenly, may not alter the average production rate, but it may carry a special message to downstream cells. The neurons which activate the muscle cells use such tricks: two impulses close together signal some muscles to double the tension produced by any following impulse.[3] Many synapses exhibit facilitation, as noted in Chapter 9, so that more neuro-transmitter molecules are released by the second impulse of a pair. When someone is tapping his foot on the floor quite regularly, and then shifts to tapping three times, pausing, tapping three times again, pausing, etc., you might well wonder if he is trying to send a message. Similarly, when a neurophysiologist observes a neuron to shift from evenly spaced impulses into a bursting mode, he asks: what caused the shift and how do the downstream cells interpret the burst? As we shall see later in Chapter 15, this question becomes particu-larly acute when observing neurons in epileptic portions of Neil's cortex. Indeed certain patterns of impulses in neurons in subcortical areas of the brain have been reported by a group of Russian investigators to signal specific words, specific groups of words with common meanings, even specific verbal decision-making.[4] Such observations, if confirmed by others, begin to indicate the impulse code which neurons use to talk to each other, which in turn, allows us to talk with each other.

What blocks the impulse in the middle of an axon? In the case of local anesthetics, the sodium channel is blocked, plugged by a molecule of the local anesthetic. How does the loss of insulating cells cause the trouble of multiple sclerosis? A special kind of cell, called a Schwann or glial cell, likes to wrap itself around the axon in between the nodes. This works much like wrapping tape around a finger so as to avoid covering the joint, and then wrapping more tape around the finger on the other side of the joint. The insulating cell

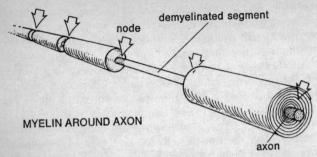

MYELIN AROUND AXON

Myelin cells like to wrap around axons, leaving only small areas exposed (nodes of Ranvier, arrows). The myelin wrapping (produced by glia or Schwann cells) is sometimes destroyed by disease (such as in multiple sclerosis), denuding 1-2 mm of axon between nodes. The impulse at a node may not be able to attract enough charge away from the next node and thus trigger another impulse; this is because the charge is more readily attracted from the naked axon near the node.

effectively makes the axon membrane many times thicker for about 1 mm of its length. When this cell is destroyed, as it is in multiple sclerosis, its insulating properties are sorely missed. The impulse at the upstream node now draws charge away from this exposed axon membrane. Naturally, the charge tends to come from nearby membrane rather than that of the next node, 1 mm distant.[5] The next node thus loses very little charge and its voltage never shifts to the threshold value. The message dies out.

This relay system is thus quite susceptible to losing messages if only 1-2 mm of the axon malfunctions; the impulses will not usually jump across more than one blocked node or more than one denuded internodal region. In the analogy, it is as if the soldiers could hear two hilltops away, but just barely. Thus, if one soldier falls asleep, the message might still get through because the next soldier can hear the gunshot faintly.

In diseases with conduction blocks, such as multiple sclerosis, there are various strategies which may temporarily improve matters. For example, lowering the temperature of the axon by several degrees may restore conduction. This is because the impulse duration increases somewhat with cooling,[6] supplying just enough additional positive charge to allow the distant node to reach impulse threshold. For complex reasons, decreases in the concentration of calcium ions (Ca^{++}) outside the axon will reduce the threshold[7]; again, this may be just enough to restore conduction in an axon which is barely failing to get an impulse across a gap. Neither strategy does anything to solve the original problem of insulation loss. They merely take advantage of our knowledge of how nerves work at the level of ions, electricity, and channels through membranes.[8]

The other thing which goes wrong with nerves, besides losing impulses along the way, is that they initiate impulses when they shouldn't. In most neurons, there is only one "right" place for an impulse to start, just as in our string of soldiers, the order to fire should be given only by the general who commands the first town. If a soldier in the middle of the chain starts firing every time he hears a door slam, a false message will be initiated. The general in the second town, hearing only the final shot at the end of the chain, assumes that it originated from the first town in the usual manner.

The human experience is full of aches and pains. Some of them are just such false alarms. Neuralgias ("nerve pains" in Latin) are pains where there is nothing wrong where it hurts: for once, the bad news really is the fault of the messenger.

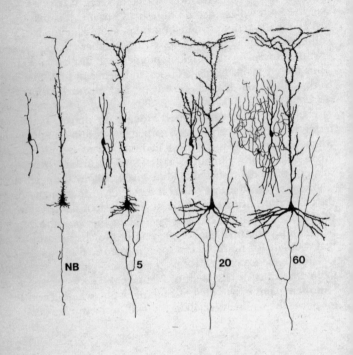

Postnatal growth occurs not only in dendritic trees but also in the arborizations of the axons. These drawings by A. B. Scheibel show the two major cell types (the tall "pyramidal" type cell and the smaller "stellate" cell) in kitten cortex near birth (NB) and at 5, 20, and 60 days after birth. Note the growth of the stellate cell's axon.

13

False Alarms: When Neurons Make Mistakes

It sometimes seems as if epilepsy operations have a cast of thousands. Dozens is actually more like it, with a half-dozen in the OR most of the time and others rotating in and out. The rest are out of sight and hearing, upstairs in the glassed-in gallery. Besides the neurologists, technicians, and engineers, there are various faculty and students who come to observe. Except when EEG decisions are actually being made by the neurologists, there is lots of talk in the soundproof gallery, talk which continues off and on throughout the whole day.

"You know, this is the kind of research that I can really get excited about," comments a third-year medical student to a professor as they watch the language cortex being mapped by the neurosurgeon. "I learned a lot in the student labs from working on frogs, but I never got excited by the research. Now this language-cortex stuff," he says, gesturing toward the patient, "you just know that it's going to be useful in treating all those people with strokes, all those kids with stuttering or dyslexia."

"Well, you must realize that this isn't really applied research," replies the professor, "it's basic research. The motivation for the research is the desire to find out how things work, to understand the underlying mechanisms. It's just like trying to understand how caterpillars turn into butterflies—except that you and I can guess better about the potential usefulness of this research than we can about caterpillar research."[1]

"I see your point," acknowledges the student. "I remember that story about Benjamin Franklin watching one of the first balloon flights in Paris with the skeptic who said, 'But of what use is it to anyone?'"

"'But sir, what is the use of a newborn babe?'" laughs the professor, anticipating the punchline. "Yes, that's one of my favorite Ben Franklin stories, too."

139

"It would really be helpful if lectures would occasionally trace through a story, starting from a basic research discovery twenty years ago, and take it through the applied research which followed, all the way up to a clinical success story today," comments the student. "Or even better, take those three levels of research, just as they are progressing today, and show how they all interrelate for one disease."

"Well, stories of the second kind are usually hard to tell without getting very technical. As it happens, I ran into a good story like that last year, while I was on sabbatical leave. I was in Jerusalem, and I remember a walk across the Old City from the home of a professor to our research lab, which was located just outside the walls of the Old City. We were commenting to each other about the diseases that you could spot in the people that you passed in the streets and alleyways."

"Sort of like being taken on a tour of the Bowery by a neurologist from Bellevue?"

"Well, not really, since alcohol-related illnesses aren't that much of a problem there. Mind you, the car drivers are suicidal, if not homicidal—but they're not drunk. No, most of the medical problems that you can spot walking the streets are either the result of poor medical care years ago, or they are war wounds. Lots of people with a missing arm or leg, bad burns, that sort of thing. Lots of chronic-pain problems. Four major wars in twenty-five years, plus the drivers and the terrorists."

The professor pours himself another cup of coffee and continues. "My friend waved at a shopkeeper along the way, a young fellow with a missing arm. He knew him because the man had a pain problem that was unusual enough to be presented at a clinic at the medical school. They have the same system that we have here, a special pain clinic where both the clinicians and the scientists from a number of disciplines get together once a week. One of the clinicians will bring in a patient whose condition has proved very hard to treat successfully. After examining the patient, the assembled experts will make some guesses at what might be responsible and what treatments might prove more successful."

"It sounds just like our pain clinic here. I was there last

week when they had a patient with a phantom limb," volunteers the student.

"Well, that's also what this man in Jerusalem suffered from— a very painful phantom. He didn't get it from a war but from an industrial accident. He was a construction worker operating a cement mixer. The jacket he was wearing got caught in the machinery and it ripped. However, the jacket stopped tearing when it got to the shoulder and began twisting his arm. It literally began flipping him head over heels, round and round, just as in one of those violent animal cartoons that people laugh at. Before anyone could shut the cement mixer off, it twisted off his arm.

"Perhaps fortunately, he was also knocked unconscious. When he regained consciousness in the hospital the next morning, the arm hurt terribly. Except, of course, there was no arm there anymore. Complete amputation at the shoulder. He said the arm felt as if it were being held behind his back, as in a wrestler's hammerlock. And this awful pain didn't go away with time, as his shoulder healed. When my friend saw him in the pain clinic conference, it was eight months after his accident. The pain was always there except that it was worse at times, feeling as if his 'arm' was being pushed upward even more forcefully."

Such a condition is called a "phantom limb" and it occurs in surgical, as well as traumatic, amputations. In a low percentage of cases, the sensation is painful. It is not a psychiatric problem. The cut end of the nerve tries to regrow (regenerate) but, in some cases, the exploring axons get lost with nowhere to go, form a tangle (called a neuroma), and become very sensitive to pressure. Impulses are initiated, which are conducted into the spinal cord; they are interpreted as coming from where they always have: the normal site of impulse initiation for those axons in the skin and muscles. This cause of phantom limb is easy to diagnose; one just presses upon the skin near the severed end of the nerve and the phantom limb seems to move, hurts more, etc. Injecting some local anesthetic at the end, or anywhere along the axon, will block the impulses and temporarily eliminate the sensations.

In some cases, such a nerve block will fail to relieve a painful phantom. This is because neurons in the spinal cord sometimes initiate false signals themselves. For example, a twisting or pulling injury to an arm may actually injure the spinal cord by pulling out the nerve by its roots; indeed, this misfortune is what happened to the man in Jerusalem. The injured cells in the spinal cord may then start responding in the same manner as a neuroma, sending signals up to the brain which are interpreted as coming from the arm.

This is but one of the possible causes of "central pain"; the patient Edith in Chapter 10 illustrates another type. Whether the abnormal impulses arise from the transducer's axon or from a higher-order neuron in the spinal cord or brain, the sensation is felt as if it arose from the skin, muscles, joints, or other internal structures of the body. The painful phantom limb is just a special case which is particularly dramatic: when the part of the body which hurts isn't there anymore.

"Well, after all this talk of phantom limbs while walking through the alleys of the Old City, we went through one of the gates in the old Turkish city walls and crossed the street to where one of the university's research labs is located. It's the old medical-school building. When the school got its new buildings completed out in the suburbs, the basic scientists were given this building. It seems to be a time-honored tradition around the world that basic scientists inherit the old cockroach-infested buildings."

"So did they start using the cockroaches as experimental animals?" asked the student.

"Right you are!" The professor smiles. "Wait a minute and I'll even tell you how the cockroach research sort of relates to the phantom-limb patient. Well, we got to the lab and the discussion shifted to some of the research in progress on rats. Now the basic question that the scientists there are interested in is one which lots of neuroscientists around the world work on—how do growing nerves recognize 'home,' how do they know where to go? It turns out that nerves compete for territory, somewhat like dogs in your neighborhood defend their own turf by barking, threatening other dogs which try to wander through 'their' yard. For example, the skin of your foot is served by

two different nerves from the lower leg, the sciatic and the saphenous nerves. There is a well-defined boundary between the skin belonging to the sciatic and that belonging to the saphenous. Suppose that you sever the sciatic nerve up in the leg. The nerve dies between the cut and the skin of the foot. So, most of the foot goes numb. But pretty soon, the numb area shrinks because the terminals of the saphenous nerve spread across the old boundary into the former territory of the sciatic."

"Nature abhors a vacuum?"

"Something like that, except that the saphenous doesn't necessarily fill in all of the vacated territory. Part of the skin may stay numb. But now the severed sciatic nerve tries to regrow, starting downward from the cut end up in the leg. When it reaches its old home, it finds some strangers living there, and a competition for territory ensues between the regrown sciatic nerve and the branches of the saphenous that have invaded the territory.[2] It is like a dog returning from the family's vacation trip to discover the neighborhood dogs have gotten used to treating the front yard as their own."

"So who wins?"

"Usually the original nerve, at least if its regrowth is not somehow hampered. Suppose the regrowing nerve gets lost and can't find its original home? For example, suppose that the foot had been amputated so that home just isn't there to be found? The end of the nerve becomes a tangle of homeless axons. Trouble is, this tangle tends to initiate nerve impulses very easily, just as do many axons when they are regrowing after being cut. That's another thing that this group of researchers investigates. Sometimes nerve impulses are produced all the time, sometimes just when you press upon the tangle."

"What does it feel like?" queries the student.

"Think about it for a minute. Where do impulses in that nerve normally come from?"

"From the foot. So does it feel just like a phantom foot?"

"Probably, although phantoms sometimes arise from more complicated causes too. Obviously, the basic research is very relevant to the clinical problem even though the major motivation for the research is actually to answer more fundamental questions: how does a nerve grow in such a way as to reach

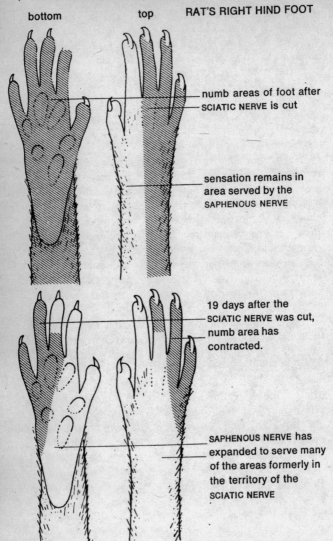

bottom   top   RAT'S RIGHT HIND FOOT

— numb areas of foot after
SCIATIC NERVE is cut

sensation remains in
area served by the
SAPHENOUS NERVE

19 days after the
SCIATIC NERVE was cut,
numb area has
contracted.

SAPHENOUS NERVE has
expanded to serve many
of the areas formerly in
the territory of the
SCIATIC NERVE

Normal territory of the saphenous nerve in rat's right hind foot (above) and expanded territory (below) 19 days after the sciatic nerve was cut. (Adapted with simplifications from Devor et al. 1979.)

'home,' how does it recognize 'home' once it gets there, how does it evict other nerves? Fundamental knowledge like that is a powerful thing. You may figure out how to relieve pain or even cure a disease by good old trial-and-error, but nothing beats actually understanding what's going on—that's where really powerful things like polio vaccines come from."

"So where do the cockroaches come in? Do they actually set traps for them under the sinks?" The student smiles.

"Ha! Actually, they have a nursery for them these days and raise their own. It helps if you know exactly how old they are and such things. In this building, there is another group of researchers whose basic interest is in how impulses travel along the nerve. Particularly how impulses fail, how they die out in the middle of an axon as if the order were canceled.[3] Lobsters and cockroaches have nice big axons which can be easily seen under a microscope, allowing you to wire them up to electrical recording equipment in much more sophisticated ways than you ever could do with the rat, which is the experimental animal used by the regrowing-nerve people. Furthermore, the lobsters and cockroaches appear to cancel out nerve impulses intentionally—they actually stop the travel of the nerve impulse, for various poorly understood reasons. But you can study the mechanism of canceling the nerve impulse, the actual electrical mechanisms which do it, because of the technical advantage offered by those big axons. Now the electrical mechanism of canceling impulses is very important information to have. Things may not work exactly the same way in humans, but most mechanisms at this fundamental electrical level have turned out to be much the same in simple animals as in higher ones. In some cases, we've had the opportunities to verify them in humans."

"So what was the connection between cockroaches and the man's phantom limb?" persists the student.

"Well, phantom limbs, like other kinds of neuralgias and paresthesias and such mischief, occur because impulses get started accidentally. From places like the regrowing nerve endings. When they reach the spinal cord and the brain, they are interpreted as coming from the 'home' of that nerve. But suppose you could cancel the impulses before they reached the spinal cord or brain?"

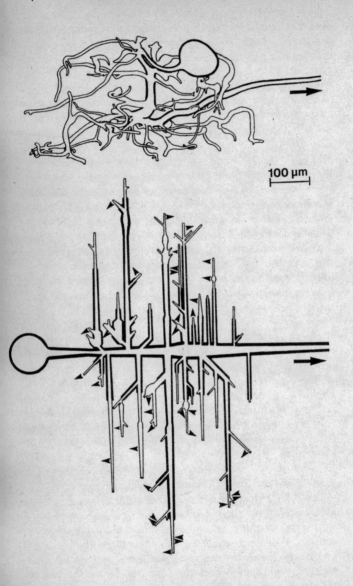

100 µm

"Nothing would be felt."

"Right. Actually, there is a suspicion that some of the impulses which start in the tangle never make it very far anyway, that they get canceled. There are several good reasons why this might happen in the regrowing axons, extrapolating from the lobster and cockroach mechanisms."

"So you might find some way of encouraging even more of the impulses coming from the tangle to fail. To eliminate the funny sensations completely," speculates the student. "It would sort of be nature's own local anesthetic."

"That's one possibility. Of course, if we could figure out how the impulses got started in the first place in the regrowing nerve endings, maybe we could prevent them rather than just canceling them while they're traveling toward the spinal cord or brain."

"So you think that the phantom-limb pains will be solved by first understanding the fundamental mechanisms in animals like lobster and cockroach, followed by trying out ideas for cures in something like a rat, and then seeing if the solution works in humans?" asked the student.

"Perhaps that's the best way to go about it, because if you understand the fundamental mechanisms, you can often illuminate lots of problems. But trial-and-error is another way of solving problems. Maybe it's less elegant and has less spinoff for solving other problems, but lots of medical problems have been solved by trying something and seeing if it works."

Invertebrate neurons have a somewhat different appearance than vertebrate neurons, but their electrical and chemical mechanisms seem quite similar. This is a lobster neuron, first drawn at top in its original shape and then again, at bottom, straightened out. The little arrows indicate some of the sites where it has synaptic connections with other neurons; the direction of the arrow indicates whether the neuron is sending or receiving. Like many of the neurons in the retina and elsewhere, these lobster neurons do not necessarily segregate their neurotransmitter release ("sending") sites onto the distant end of an axon. Thirty such neurons are contained in the stomatogastric ganglion of the lobster and run the chewing and squeezing actions of the stomach (lobsters have teeth in their stomachs). Analysis of this small group of neurons has been very important for understanding how neurons work together in a group (drawings from King 1976).

Neuralgias do not require something as dramatic as twisting an arm off for their initiation, or even cutting a nerve. A typical neuralgia is one blamed upon a "slipped disc": an aching pain is felt in the leg and foot, becoming worse with movements. In many cases, this is due to a disc pinching the nerve as it enters the backbone on its way to the spinal cord.[4] Impulses are initiated at the pinched region between two vertebrae, but they are interpreted as coming from the leg and foot, since that is where impulses in those axons normally come from.

This may seem simple and straightforward, but it has puzzled neurophysiologists for decades, ever since they learned

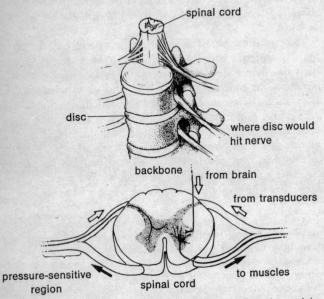

The spinal cord inside the backbone (vertebral column), receiving the axons of transducer neurons (open arrows at sides) and sending out axons from motor neurons to run the muscles (filled arrows). The motor neurons also get instructions from the brain via axons descending in the white matter. Where the nerve passes through the bone, it is near the intervertebral disc. When the disc herniates, it may press upon the nerve and initiate impulses which the brain interprets as coming from the skin and muscles.

how hard it is to start up impulses in a nerve by merely pinching it. Normal nerve has several protective mechanisms which keep it from initiating impulses on its own. First of all, the middle of an axon has somewhat limited abilities to change its voltage when distorted, compared to the region of the neuron in the skin which is specialized for detecting mechanical stimuli. But even when such voltage changes do occur in the middle of an axon, they are usually ineffective at initiating an impulse because a node requires a sudden voltage change. A voltage which builds up slowly causes the sodium channels to close, in the same manner as they do at the peak of an impulse. Thus the puff of sodium ions into the cell is prevented. This accommodation characteristic means that an impulse will not be initiated unless the voltage change is rapid—as it is, of course, when an impulse occurs at the adjacent node just upstream. Regions of neurons which normally initiate the impulse, such as the beginning of the axon, lack such accommodation characteristics, besides having greater sensitivity to mechanical distortion.

It seems that chronic injury to a nerve is different from the experimental-model situation which neurophysiologists have studied since 1929; pinching a normal nerve usually fails to initiate impulses, but pinching a nerve which has been irritated for a few days by a foreign body will easily yield impulses. It doesn't even require a sharp pinch: impulses are initiated from such a chronically irritated nerve just by stretching the nerve slightly, in the manner that it is stretched by standing up and walking around. And, to come full circle, one of the responses to such chronic irritation is that the insulating cells die and expose the nerve.

There are, alas, many kinds of neuralgias.[5] Common to many of them is loss of the protective mechanisms discussed earlier: although normal axon requires rapidly changing voltages, injured and regrowing axons not only produce local voltages in response to distortions but can respond to such slow voltages with many impulses.

How does the change in axon properties occur? What does the loss of insulating cells have to do with it? What about the piercing pain in Betty's face when her lip was lightly touched (tic douloureux)? Sorry, few answers yet.

FRONTAL LOBE

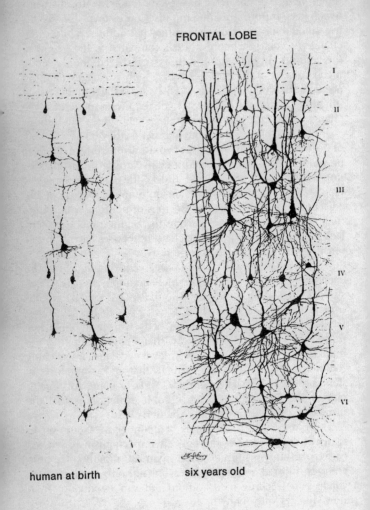

human at birth   six years old

These Conel drawings are from the same region of frontal lobe as that compressed by Tom's tumor (see next figure).

14

Emotions in the Brain: Frontal Lobes, Personality, Depression, and Psychosurgery

Tom had always been outgoing, a joiner, president of his local club, a leader of a scout troop. Everyone had thought of him as a true extrovert. In retrospect, his wife noticed some changes beginning a few years before. He became quieter, didn't go out much, seemed to care less about what went on around him. "But we're all getting older," she thought. Later, he sat around not doing anything and not saying very much. After many months of this, his wife finally persuaded him to see a psychiatrist. The psychiatrist told her that she should take him to a neurologist, that Tom's behavioral change looked "organic."

The neurologist noted only a few unusual things in her examination of Tom: when she asked him the meaning of the proverb "People who live in glass houses shouldn't throw stones," he just repeated the sentence back to her. When she asked him for more of an explanation, he just repeated it back again. He didn't say much spontaneously, and he didn't show any reaction later when told that he was seriously ill and needed an operation. Although memory, language, motor, sensory, and coordination tests were all normal, his eye exam showed changes. Using an ophthalmoscope, the neurologist could see the vessels which come from the brain to supply the eyes with blood. They showed signs of pressure building up inside the skull. Later, a computerized tomographic scan showed a large tumor growing in the meninges, the membranes which cover the surface of the brain. The tumor was greatly indenting both frontal lobes. That type of tumor is benign; it grows relatively slowly and locally, producing symptoms only because it compresses the underlying brain. If it is in a place where surgical removal can

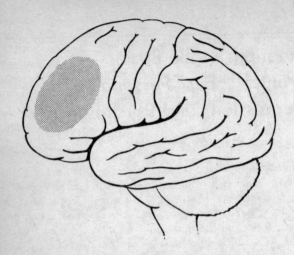

Tom's tumor, pressing upon the frontal lobe.

be accomplished, as it was in Tom's case, the patient can be cured. A year later, Tom was his old outgoing self.

The biggest difference between the brains of higher primates and that of man is in the size of the frontal lobes, yet we know less about the frontal lobes than any other area of the brain. Damage to both frontal lobes seems to alter two things: the ability to alter behavior with changing experiences, and the degree of emotional responsiveness.

If Tom had been tested earlier in his illness with a special neuropsychological test battery called the Wisconsin card-sorting task, he would have shown a peculiar deficit. This task requires the subject to sort a deck of cards into two piles; for example, all face cards in the right pile, all others in the left pile. [1] But the subject is not given that specific instruction; the subject is simply supposed to start sorting cards into the two piles and the examiner will say "yes" or "no" after each card (depending upon whether a face card was placed in the right pile, other cards in the left pile). Subjects catch on to the criterion after a short while.

Then halfway through the deck of cards, the examiner changes the criterion, saying "yes" when red cards are placed in the right pile, black cards in the left pile. Of course, the examiner doesn't mention that the criterion has changed. A normal person catches on that the rules have changed and figures out the new strategy. A person with frontal-lobe damage will usually simply persevere with the first strategy. With more extensive frontal-lobe damage, the ability to abstract is lost (as in Tom's response to the proverb). Eventually, the patient may simply repeat actions (or words) over and over, a phenomenon called "perseveration."

In addition to this loss of flexibility in coping with changes in the environment, patients with frontal-lobe damage seem to lose emotional responsiveness, becoming "flat," without happiness or grief, seemingly little concerned with the effect of the environment on them, or with the effect of their own behavior upon others. Scolding the patient for urinating on the living-room floor may have little effect.

Although the mechanisms in the brain underlying emotion are still not entirely clear, some features are known. The areas of brain where damage alters emotional responsiveness seem to be the same brain sites concerned with visceral function: regulation of the heart rate, blood pressure, respiration, digestive activity, and the levels of various hormones. There is indeed a relation between emotion and "butterflies in the stomach." This overlap of brain sites for emotion and visceral function suggests an anatomic basis for "psychosomatic" diseases.

Visceral regulation in man depends upon two systems. One, the sympathetic nervous system, prepares the body for "fight or flight," raising heart rate and blood pressure, erecting the hair, decreasing digestive activity. The other, the parasympathetic nervous system, does the reverse, preparing the body for more vegetative activities. The head offices for these two systems are located in the hypothalamus (below the thalamus, above the pituitary gland); the hypothalamus also regulates hormones via the pituitary. Damaging the hypothalamus in experimental animals alters their emotional responsiveness. When a small area of the outer part of the hypothalamus is destroyed, cats will change from being easily handled to responding with "blind rage" to all contact. Cats with damage a few millimeters away, in inner parts of the hypothalamus, never show rage; they become placid and fat. There are a few human cases, with tumors or small strokes in the hypothalamus, where such syndromes have been seen. There are also areas in the hypothalamus and associated pathways where electrical stimulation seems to be quite pleasurable: given a choice of food, water, sex, and a bar to press which will activate the stimulator, animals will press the bar until exhausted.

At this hypothalamic level, emotions of rage and pleasure are little altered by any environmental stimuli; it is at the cortical level that the interaction between environment and emotions seems to occur. Again, it is at those sites in the cortex where visceral functions can also be altered, such as the inner surfaces of the frontal lobes and temporal lobes. These areas have anatomical connections with the hypothalamus; together they are called the limbic system.[2] Extensive damage to the temporal-lobe portion of the limbic system can lead to an animal (or man) who is more likely to respond to any environmental input with rage. Damage to the frontal-lobe portions of the limbic system can lead to placidity and indifference, as in Tom's case.

Wide areas of cortex seem to interact with the limbic system to bring about that mixture of emotional character and thought which we label personality. There are some differences between the two hemispheres. For example, an epileptic focus in the right temporal lobe is more likely to be associated with a personality that shows more emotional responsiveness, sexuality, concern for detail, and helplessness. Such patients also tend to underrate their deficiencies and overrate their own abilities; they "polish their image." A left-temporal-lobe focus, on the other hand, is more often associated with a moralistic, religious, rigid, sober, self-deprecating personality, tending to emphasize the personal significance of events (often writing down all this personal significance in great detail).[3] These are, of course, simply average tendencies: Neil's personality does not fit the left-temporal-lobe "type."

The right brain seems in general to have more to do with emotional experience and perception than the left brain. The

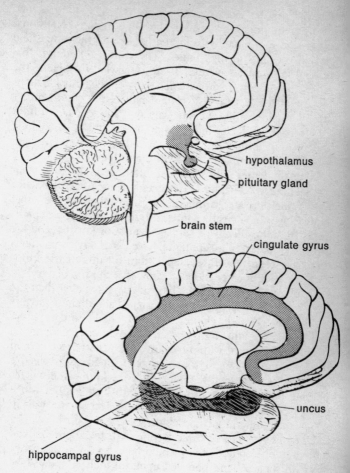

The left brain, seen after a cut down the midline. In the bottom drawing, the brain stem and cerebellum (seen in the top drawing) have been removed so that the underside of the left temporal lobe can be seen. The dotted areas are the limbic system, concerned with emotion. The cingulate gyrus is the frontal-lobe portion of this limbic system; the uncus and the hippocampal gyrus are its temporal-lobe parts. The hippocampus itself is just beneath the hippocampal gyrus.

left half of the face (controlled by the right brain) shows greater emotional expression, but one notices emotional expression more readily using the left half of the visual field (which goes first to right brain). This means, of course, that normal face-to-face encounters are somewhat inefficient: the more emotionally perceptive left half of the visual field is looking at the less expressive right half of the other person's face. (Couples could, of course, always stand together in front of a mirror while having a domestic argument.) Laughing and smiling, unlike other sequential facial movements, are more readily disrupted from right brain than left.[4]

Is there damage to particular brain areas in the major mental illnesses? That has been difficult to discover. A tendency toward such diseases is inherited. This strongly suggests that something is biologically wrong with the brain in these conditions. Both the affective disorders (depression, mania) and the thought disorders (schizophrenia) have hereditary components that have clearly been separated from environmental influence. Studies of early adoption show that the risk of developing a psychosis goes with the biologic parents rather than the adoptive parents. Identical twins separated at birth have roughly similar chances of developing a major mental disorder.[5] In an earlier chapter, the hint that some cases of schizophrenia might be associated with an excess of dopamine receptors in the striatum (a part of the brain concerned with selective attention) was presented. Other evidence from some schizophrenic patients, who also happened to have temporal-lobe epilepsy, has implicated a congenital defect in the way in which the temporal lobe develops.

One observes more people having both schizophrenia and temporal-lobe epilepsy than would be expected on a chance basis. In such patients operated upon for their temporal-lobe epilepsy, an unusual kind of anatomy is revealed: little pockets of gray matter buried in the white matter of the temporal lobe.[6] Unfortunately, their removal, though helping the epilepsy, doesn't help the schizophrenia much. Auditory hallucinations also occur in both diseases. The evidence is thin, but it is tempting to think that altered function of the temporal lobe has something to do with schizophrenia.

There is better evidence for abnormalities in neurotransmitters, as we discussed earlier. In the same manner that dopamine is suspected of a key role in schizophrenia, so norepinephrine and serotonin are suspected of a role in profound depressive disorders.[7] Drugs that increase the levels of those neurotransmitters, such as the tricyclic antidepressant medications, have proved most useful in the treatment of many cases of depression. But there remain a few patients whose depression persists despite this drug therapy.

Edmund was an honor student in his final two years of high school. His parents didn't notice anything wrong until Edmund tried to kill himself. Edmund himself had noticed the change months before. Schoolwork wasn't worth doing. Tennis wasn't fun—he dropped out of that with an excuse about a knee injury. He didn't have enough energy for anything. He felt that he wasn't good enough for his girl friend; he saw her less and less. By the time that he took all of those sleeping pills, he was sure that his life just wasn't worth anything.

The psychiatrist started treating the symptoms of clinical depression right away after Edmund recovered from the overdose; he started Edmund on one of the new antidepression medications. But that didn't change anything except to make Edmund sleepier. There were therapy sessions with his parents and brother and sister. His family tried to help him in every way, but Edmund felt that he wasn't worthy of them. Nothing changed that awful feeling of blackness. Some days were better. For a week, he was out of the hospital, but the depression returned. He was then back in the psychiatric hospital, with more therapy and different medications.

Later, the psychiatrists recommended a series of electroshock treatments, since everything else had failed. Years ago, it was noticed that psychotic patients who also had epilepsy seemed less psychotic after they had a big seizure. Artificially inducing seizures with electric shock (using paralyzing drugs to help protect the patient from injury) proved to help many, especially those with severe depression. But Edmund felt no better after a series of electroshock treatments; the only way to get rid of that awful blackness, that pain and unworthiness, was to die. While being walked around the hospital grounds by an orderly, Edmund broke away and jumped in front of a car. The result was a broken leg and more therapy sessions, and more drugs, and more electroshock. Still, nothing helped: he was still empty and rotten inside, not worthy to live. Suicide was the only answer.

Most patients with severe depression respond to treatment with medications and psychotherapy. The exceptions usually respond to electroshock treatments. But that is small comfort to Edmund, who happens to be one of the smaller number whose symptoms persist after the standard treatments. If Edmund were in Great Britain, the next step would be referral to a center for psychosurgery. Several British centers have reported on their extensive experience with operations which destroy a small portion of the inner side of both frontal lobes, in part of the limbic system.[8] This apparently provides complete relief from depression in about half of the patients like Edmund; another third of the patients are improved, although they do not get full relief. In the British experience, the improvement in symptoms of depression lasts for at least five years or longer, with side effects in less than 2 percent of the cases. Measurable personality changes have been rare. Scores on formal psychological tests, including IQ tests, generally are either unchanged or improve after the operation.

Despite this favorable British experience with modern psychosurgery, Edmund will probably not be treated with such an operation in the United States, even though there would seem to be no other options except waiting (or suicide). In some states, the operation can be legally done only upon approval of a state board, a committee of physicians and laymen who will never see Edmund. That committee must give its approval even if Edmund and his family want the operation. Such committees sometimes find it easier to say no; permitting the operation would mean sharing responsibility in case something went wrong. Elsewhere in the United States, the operations may be considered too much of a political hot potato by the psychiatrists and neurosurgeons, not worth all the trouble for the few patients whom they might benefit. At most, only a few hundred of these operations are done yearly in the United States.

The history of "prefrontal lobotomies" has a great deal to

do with the political controversy over the modern psychosurgical operations. The original operation was controversial in its day, even though it marked a major advance in the treatment of the mentally ill. Animal studies had indicated that cutting the connections between the very front of the frontal lobe ("prefrontal") and the rest of the brain was associated with more docility and placidity. These results were presented at an international physiological meeting in 1935, where they came to the attention of an inventive Portuguese neurologist, Antonio Caetano de Abreu Friere Egas Moniz. Based upon these animal experiments, he devised the operation of prefrontal lobotomy, where the same connections were cut in man.

This innovation had a number of effects: Moniz won the Nobel Prize; his medical practice was disrupted by political demonstrators opposed to psychosurgery; and, for the first time, a treatment was available that allowed some of the severely mentally ill to leave the asylum and return to society.

It was soon recognized that this was an imperfect treatment. The operation was largely abandoned in favor of the psychotropic drugs when they first became available in the 1950s. Although many of the lobotomized patients had improved enough so that they could leave the asylum, they often demonstrated some personality changes. Like Tom, who had the tumor pressing upon the frontal lobes, some lobotomized patients tended to be uninterested in their surroundings, or in the effects of their own behavior upon others. Patients with thought disorders (schizophrenia) seemed to be helped less by prefrontal lobotomy than patients with depression, phobia, and obsessions.

Essentially, the prefrontal lobotomy was too large an operation; a small area of destruction works as well, as subsequent surgeons discovered, and then there are far fewer undesirable side effects such as personality changes. In the modern operations, only selected areas on the inner side of the frontal lobe are destroyed, using x-ray placement of probes as in the operation on the thalamus for Parkinson's disease. Then the "uninterested" personality change does not occur and great improvement is seen in depression, phobia, and obsessive behavior (again, schizophrenia is not helped as much by these operations). Behavioral improvements result only if the right location is destroyed, and not if the lesion is made nearby. There is something specific about that portion of the brain in changing the symptoms of depression.

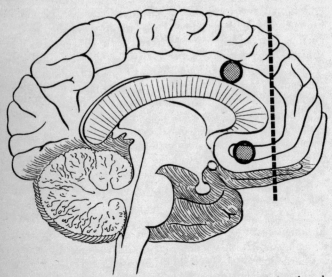

A summary of psychosurgical operations, both outmoded and modern. The left brain is seen from the midline, with the frontal lobe on the right of the drawing. The dashed line is the cut made in the original "prefrontal lobotomy," an operation which has not been performed for many years. The dotted circles show the sites of several of the modern psychosurgical operations; usually, only one of these targets is destroyed in a given patient. The upper target is for an operation called a "cingulotomy"; the lower, an "orbital-frontal tractotomy."

The debate over psychosurgery has been clouded by the failure to distinguish between the results of modern operations, with their more specific indications and fewer side effects, and the original prefrontal lobotomy. This is rather like debating airline safety without distinguishing between the airplanes of 1936 and 1980. With the development of more effective drug treatments in psychiatry, the number of patients with persisting illness has also decreased so that surgery is now indicated in only a very occasional patient with depression, perhaps a few in a thousand. For that rare patient such as Edmund, however, the operation may be lifesaving; unsuccessfully treated profound depression is a disease with as high a mortality rate as cancer.

For many people, psychosurgery stirs up a negative reaction: "It just isn't natural," or "I wouldn't want anyone messing around inside *my* head." Similar reactions often occur when science touches upon other aspects of emotions or sexuality, and such responses are presumably "natural" in the sense that they flow out of individual experiences and genetic heritage. The basic problem is not the reactions but what often follows: the failure to examine evidence and to discuss alternatives. The enshrinement of the "natural" over the "artificial" usually ignores the fact that nothing is more natural than suffering and death, poverty and disease. Jonathan Miller, in *The Body in Question*, notes that "far from rejecting or distorting nature, scientific medicine achieves its results by recognizing what nature is and by reproducing and reconstituting her grand designs. Science is not a blasphemy; the willful rejection of its insights is."

CINGULATE GYRUS

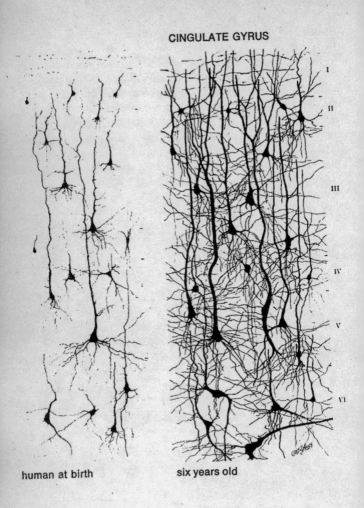

human at birth   six years old

The cingulate gyrus is one of the targets in the modern psychosurgical operation called the "cingulotomy" (see Chapter 14).

15

Inside the Epileptic Focus: Sick Neurons, Sick Circuits, and Scars

It is now afternoon in the OR. The neurosurgeon, hearing that a nurse is going down to the cafeteria, asks if a hamburger might be brought back for him. Neil calls out an order for pastrami on rye—in jest, as he knows that all he is likely to get is glucose, intravenously. Anesthesiologists have a well-founded dislike for full stomachs in their patients.

A new OR crew is beginning to appear. Several new doctors scrub in. One is another neurosurgeon, who will take over while his colleague takes a break. Neil's neurosurgeon needs this break not only to sit down outside and eat his hamburger, but also to think about what he has learned, to plan how much temporal lobe to remove, to think about the problems of the epileptic focus bordering upon Neil's language cortex.

The other doctor is a neurophysiologist, a scientist who spends most of his time studying how the individual nerve cells of cats or lobsters use electricity to signal one another. He will assist the neurosurgeons in recording the electrical activity of individual nerve cells in the epileptic focus. Techniques developed in the research laboratory are transplanted to the OR; most of the required amplifiers, stimulators, and tape recorders which characterize a modern brain-research laboratory are located upstairs in the gallery. A sterilized box of special electronics equipment is unwrapped and plugged in. A very fine needle, covered with electrical insulation except at its very tip, is used to record electrical signals from individual nerve cells. The individual nerve cells cannot be seen because they are packed tightly together, so the fine needle is lowered blindly into the epileptic focus until a neuron is randomly encountered, much like fishing. Amplified electrical signals from the neuron are

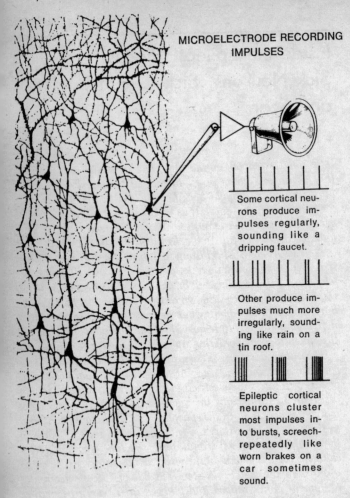

MICROELECTRODE RECORDING
IMPULSES

Some cortical neurons produce impulses regularly, sounding like a dripping faucet.

Other produce impulses much more irregularly, sounding like rain on a tin roof.

Epileptic cortical neurons cluster most impulses into bursts, screech-repeatedly like worn brakes on a car sometimes sound.

then displayed on the large TV screen and also played through a loudspeaker.

Soon the popping sound of a nerve cell is heard throughout the OR. Neil is fascinated, becoming one of the few humans *who has been able to listen in on the electrical conversations of one of his own brain cells. Each impulse makes a pop in the loudspeaker, and the pattern of impulses is thus easily heard. Some nerve cells produce a rhythmic train of impulses, sounding like a dripping faucet. Other nerve cells are quite irregular, sounding more like a light rain on a tin roof. But many nerve cells in the epileptic focus tend to produce impulses furiously, so that the impulses cluster. This creates a sound variously described as a screech or as a Bronx cheer. Over and over these "epileptic bursts" occur. Is this an imperative signal, compelling all downstream nerve cells to pay attention?*

The human nerve cells seem to operate on the same electrical basis as those in monkeys, cats, or even lobsters, as far as one can tell. The nerve impulses and other electrical signals are produced by moving sodium, potassium, chloride, and calcium ions across the cell's surface membrane. For most purposes involving the study of electrical operation of nerve cells, animal nerve cells will substitute nicely. But this can be applied only to the normal operation of nerve cells, as we know rather little about the abnormal behavior of nerve cells in most animals.

One of the most important first steps in understanding a disease is to be technically able to induce the disease in a laboratory animal. This multiplies the opportunities to study the disease, and in a way that permits experimental conditions to be controlled, potentially harmful treatments to be tried out. Even producing a cross-eyed monkey can be difficult: monkeys often successfully compensate for the surgical shortening of one of the muscles attached to the eye. But how does one produce epilepsy?

There are various experimental ways of making nerve cells sick, such as applying toxic chemicals or producing a scar. Some methods even produce seizures. But is this epilepsy? They are not good models of the human disease "epilepsy" unless they involve chronic changes in nerve cells, the kind that lead to the characteristic abnormalities of the EEG day in and day out, even when a seizure is not imminent. How does one keep animal research on the right track, in the attempt to induce true epilepsy in lab animals and thus multiply the opportunities to study its mechanisms? One essential is to constantly compare animal results with the real thing: analogous data from human epileptic brains.

The neurons in an artificially epileptic animal brain may screech away in a manner superficially similar to that of many of the neurons seen in the human epileptic focus. A more detailed computer-assisted examination of the impulse timing, however, reveals some major differences between the human neurons and those in most animal models for epilepsy. Only one of the epilepsy models, where a metal is used to slowly produce a permanent scar in the animal's cerebral cortex, produces nerve impulses timed in the manner of the human epileptic cells.

Upstairs in one of the research laboratories, a monkey is looking at a meter. He has learned from experience that whenever the meter swings to the right far enough, a squirt of apple sauce will come from a tube in front of his mouth. The monkey is making one of the nerve cells in his brain produce nerve impulses. A computer counts these nerve impulses and, when enough occur in the right pattern, it triggers the squirt of apple sauce. The computer also runs the meter's needle, telling the monkey that he is getting closer to or farther away from the goal. The researchers, in turn, tell the computer what is "good" and "bad." If they are smart enough, monkeys can learn to increase the rate of nerve impulses from one nerve cell, while simultaneously decreasing those produced by an adjacent nerve cell in the monkey's motor strip.

The monkey is sitting inside a soundproof room, watching the meter swing and licking up the apple sauce every few seconds. A small chamber is fixed atop a hole in the monkey's head, created months before under sterile conditions and general anesthesia. Every day that the monkey comes to work, a remotely controlled device is attached to this chamber. A very fine needle, just like the ones used in the OR, is advanced slowly and carefully through the motor strip by remote control. The neurophysiologist listens to a loudspeaker for the sound of a nerve cell. When this "fishing" yields a nerve cell, he halts

Neil is asleep. Neuron in
epileptic focus fires in bursts.

Neurosurgeon speaks loudly,
"Neil? Wake Up!"

Neil responds, "Huh?"

Neil is now awake.

Even though Neil is awake,
bursts continue unchanged.

One minute later.

Electrical discharges from a single neuron in human temporal lobe, prior to the removal of the epileptic focus. These recordings were made while the patient was napping; in the second panel, the neurosurgeon tries to wake up the patient. Several seconds later, the patient responds, "Huh?" The last three panels show the neuron's electrical activity while the patient is awake. Each frame shows 0.5 second of activity. Unlike most normal neurons, the epileptic neuron clusters its impulses into bursts, with only several milliseconds between each impulse (500/sec firing rates). This neuron also shows relatively little change from sleep to waking, unlike most normal neurons. Adapted from Calvin, Ojemann, and Ward, *Electroencephalography and Clinical Neurophysiology* 34:343, 1973.

the needle and the electrical activity from the nerve cell is observed, sometimes for many hours.

This particular rhesus monkey is an epileptic, made artificially epileptic by a scarred region of brain in the left motor strip. When he has seizures, they begin with his right hand starting to twitch, followed by his arm, and sometimes the whole right side of his body. Many of his nerve cells in the motor strip screech away in an abnormal manner, even when he is not having a seizure but is just sitting there, working at getting more apple sauce. Today, the computer which runs the apple-sauce dispenser is programmed to examine the nerve impulses for the screeching "epileptic" patterns. They are considered "bad" and make the meter move left. The "rain on the roof" irregular pattern of nerve impulses, which normal nerve cells exhibit, is considered "good" today and runs the meter to the right, toward the next squirt of apple sauce.

Monkeys can actually learn, somehow, to control the impulses from such neurons, forcing them away from the abnormal "epileptic bursts" and causing them to act more like normal neurons. The monkeys just require some feedback about whether they are getting closer to or farther away from the goal (watching the meter) and some incentive (the apple-sauce dispenser). As a lab visitor remarks, "What a way for a monkey to earn a living!"

If a single nerve cell dies, it will never be noticed. Indeed, it was once estimated that over 10,000 nerve cells die every day in a human brain; unlike the other organs of the body, the brain does not make new cells, so the total numbers are always declining during one's lifetime.

If the "epileptic" nerve cells would simply die, they would probably cause little trouble. It is the chronically sick nerve cell which causes so much trouble. They incite others to riot, as it were, and occasionally produce seizures.

Basic studies of nerve cells, those directed at understanding how nerve cells work rather than aimed at curing a particular disease, have provided much insight into how bursts of impulses are generated normally. Hormones, for example, can change the rhythmic impulse patterns of a snail nerve cell into the bursting patterns. In many types of nerve cells,[1] there is a "two impulses for the price of one" mechanism, which sometimes gets carried away and produces "ten for the price of one." When a cell loses many of its connections with upstream nerve cells, it reacts in unusual ways to this loss. It may begin producing impulses even with no inputs to trigger them. It may overreact when inputs do occur.

Neil's brain indeed has many nerve cells which screech away, over and over. Each time that the recording needle is inserted into a new area of temporal lobe, a few nerve cells are encountered. One surprise from studying the human epileptic brain is that there are so many normally acting nerve cells, mixed in with those which act strangely. Of course, when a seizure is imminent, these normally acting cells also begin to produce impulses in a rapid-fire manner.

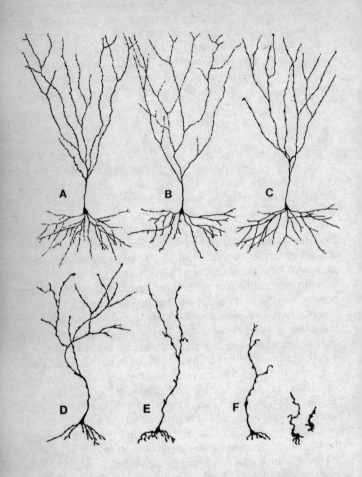

In temporal-lobe epilepsy, some neurons in the hippocampus may atrophy. Normal neurons (A) are seen, as well as neurons in a number of intermediate stages of atrophy (from M. E. Scheibel, P. H. Crandall, and A. B. Scheibel, "The hippocampal-dentate complex in temporal lobe epilepsy. A Golgi study," *Epilepsia* 15:55-80, 1974; reprinted with the permission of the authors and of Raven Press).

But between seizures, only some of the neurons in an epileptic focus act strangely. If only one could eliminate those nerve cells selectively, instead of having to surgically amputate so much of the temporal lobe. Better yet, we need to cure those pathological nerve cells and let them live, contributing to whatever brain function they perform.

TIP OF TEMPORAL LOBE

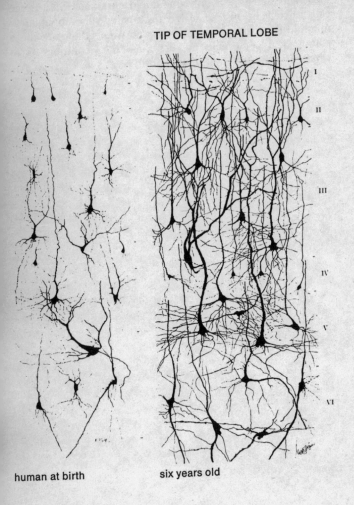

human at birth   six years old

16

Decisions and Dilemmas: Removing the Focus

Lunch is over for the neurosurgeon. He once again scrubs in and joins his colleagues standing around Neil's head to discuss his plan for how much brain to remove. If he removes too little, the seizures may be unaffected. But removing too much near the rear of the temporal lobe may affect Neil's language abilities, one of the things which makes him uniquely human and also uniquely Neil.

The numbered paper tags lying on the surface of Neil's brain serve to remind everyone what's where. The spots where the epileptic waves in the EEG were most prominent are marked with special paper tags having letters on them rather than numbers. Using sterile paper and a sterilized pencil, the neurosurgeon has made a list of this and a drawing to keep track of all the locations and test results.

A compromise plan is arrived at. The front portion of the temporal lobe will first be removed, leaving the rear portion intact for the moment. Finally, in the middle of the afternoon, the neurosurgeon begins removing the epileptic focus. Neuroscientists who study the anatomy and neurochemistry of epilepsy will use specimens of this epileptic brain to try to figure out what is abnormal about it. In order to know what to look for, they are relying on the understanding of normal brain structure and function which has been built up over the years by basic researchers. The differences are often subtle in the epileptic brain; only occasionally will an obvious problem, such as a tumor, be responsible.

After an hour of careful surgery, the front part of Neil's left temporal lobe has been removed. Neil did not feel anything. He napped on and off throughout the process, occasionally asking how things were going and getting progress reports. Now that the neurosurgeon has finished the removal of brain tissue, he asks that the neurologists be paged and begins setting up the EEG recording equipment again.

The EEG experts arrive upstairs in the gallery shortly thereafter, and the TV monitor in the OR again displays Neil's brain waves, especially those from the rear of the temporal lobe. There is a considerable effort to make sure the electrical recordings are technically perfect, as the big question is whether or not epileptic EEG patterns remain in the rear of the temporal lobe. The neurosurgeon has left a few centimeters (more than an inch) of temporal lobe in front of where Neil's language area seems to begin, and would prefer not to come any closer unless the epileptic activity is severe.

To everyone's disappointment, there are some epileptic-type voltage patterns remaining on the EEG from this strip of temporal lobe. A long conversation ensues via the intercom linking the neurologists in the gallery with the neurosurgeon in the OR. How much is too much? Finally, the neurosurgeon decides to go no further, hoping that the remaining epileptic tissue will be insufficient to evoke a seizure from the rest of the brain. But only time will tell.

It wasn't very many decades ago that the only thing that could be done for the 1 percent of humans who were epileptic was "supportive care." Many epileptics could not work. Many of the more disabled epileptics required supportive care in an institution. The public's understanding of the brain was much poorer then than now; in earlier centuries, epileptics were even burned as witches, "possessed by the devil." An occasional epileptic was inspired or energized by persuasive hallucinations. Paul on the road to Damascus? Joan of Arc hearing voices? But for most victims, epilepsy is a disorganizing and disruptive aspect of their lives which others fail to understand.

The biggest change came with the development of effective drugs. Current medications are now effective for about 75 percent of epileptics, suppressing most seizures even if not curing the disease. But there are many epileptics for whom even modern anticonvulsant drugs will not work effectively. About 10 percent of this group may benefit from surgery, as they, like Neil, happen to have an epileptic focus which is located in a portion of the brain which can be removed

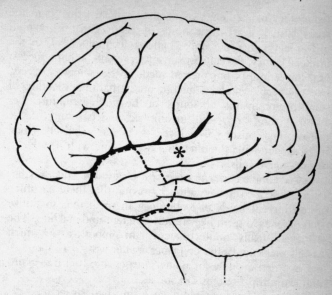

The amount of brain removed from Neil in the epilepsy operation; asterisk marks the site of the remaining epileptic activity in the EEG afterwards.

without the cure being worse than the disease. But there are many epileptics left in the predicament of having no effective treatment available. These people are dependent upon the pace of medical research, hoping for a new drug or a new operation. But what they are really dependent upon is how fast researchers can overcome our general ignorance of the brain. Epilepsy is merely a symptom which arises from a variety of conditions which damage the brain: head injuries, congenital malformation of the brain or its blood vessels, tumors, some toxic substances, or a shortage of oxygen or glucose in the blood supplying the brain. Sometimes one can fix a sick TV set by kicking it, but a detailed knowledge of the principles on which the TV set operates is more likely to be effective.

The epilepsy surgery, with all of its attendant technology providing electrical monitoring of brain activity and electrical stimulation to modify brain functions, seems like the high technology of medicine, as do the artificial kidney and the transplanted heart. But none of them constitute "high technology"; they are only "halfway technology." Like the artificial kidney, the epilepsy surgery represents both the success and the failure of our present medical technology. Success, because people can be helped by our partial understanding of how to circumvent the source of the trouble. Failure, because it shows how our incomplete understanding of the disease mechanism has reduced us to one relatively crude method: removing the offending tissue—and with it, a lot of normal nerve cells.

Lewis Thomas, in *The Lives of a Cell* (New York: Viking, 1974), described the situation by saying that there are three quite different levels of technology in medicine, so unlike each other as to seem like entirely different undertakings. The first level is really "nontechnology," the supportive care which was all one could offer epileptics in the past. It is about all that one can do now for many diseases affecting the brain, such as multiple sclerosis or strokes.

The second level of technology is the "halfway technology" which so attracts public notice: building a machine to mimic the kidney's functions, transplanting a heart to replace a diseased one. Like supportive care, these halfway technologies are very expensive. A year's care on an artificial kidney runs between $15,000 and $25,000, all because we do not yet understand the immunological process attacking the kidney cells which causes the kidney failure in the first place. Epilepsy operations are expensive because many dozens of professionals are required to make them possible. The one hour of actual tissue removal is only a minor portion of the ten to twelve hours in the OR. The day of surgery is only a minor portion of the many months of preoperative and postoperative care by many professionals which go into the handling of a difficult seizure problem.

Advances in this second level of medical technology account for much of the rapid rise in the costs of health care. It is this second level of technology which must be limited to contain those costs, whether through federal legislation or hospital utility commissions. Some have advocated a limit on the rate of growth, or an absolute limit on health-care expenditures as a percentage of the gross national product. But who is to determine how to ration out this second level of technology? Is it the auto worker with chronic kidney failure who will be denied a kidney machine, or the executive with coronary-artery disease who will be denied a coronary-bypass operation, or perhaps the epileptic like Neil who will be denied his operation? Limits on health-care expenditures mean limits on these expensive halfway technologies. One hears a lot about limits on health-care expenditures. Not much is heard about how such technology is to be allocated. That touches us all, and needs to become a part of the public discussion.

The third level of technology is so effective, Thomas notes, that it seems to attract the least public notice; it has come to be taken for granted. "This is the genuinely decisive technology of modern medicine, exemplified best by modern methods of immunization . . . and the contemporary use of antibiotics and chemotherapy for bacterial infections." Just consider how the polio vaccines have bypassed the iron-lung technology. He continues, "The real point to be made about this kind of technology—the real high technology of medicine—is that it comes as the result of a genuine understanding of disease mechanisms, and when it becomes available, it is relatively inexpensive and relatively easy to deliver." One can look forward to a day when epilepsy operations are seldom necessary because we understand enough about the underlying normal mechanisms so as to work around the problems presented by disease.

"It is when physicians are bogged down by their incomplete technologies, by the innumerable things which they are obligated to do in medicine when they lack a clear understanding of disease mechanisms, that deficiencies of the health-care system are most conspicuous. If I were a policymaker, interested in saving money for health care over the long haul, I would regard it as an act of high prudence to give high priority to a lot more basic research in biologic science. This is the only way to get the full mileage that biology owes to the science of medicine, even though it seems, as used to be said in the days when the phrase still had some meaning, like asking for the moon."

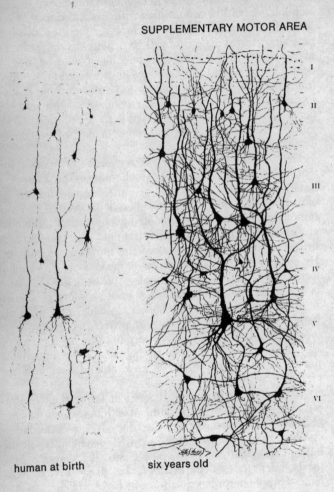

SUPPLEMENTARY MOTOR AREA

I
II
III
IV
V
VI

human at birth   six years old

17

The Brain Understanding Itself: Epilogue and Prologue

The rest of the epilepsy operation is anticlimax. The working day is over for Neil, and he gets the intravenous equivalent of a sleeping pill so that he can sleep through the closing-up stage of the surgery. After color pictures are taken, the remaining paper tags are removed from the surface of the brain. The scrub nurse counts them to make sure that none are missing. Then the flap of dura is folded back and sewn in place. The scrub nurse unwraps the large piece of bone which was removed from the skull and the neurosurgeon anchors it back in place over the opening. The scalp flap is sewn back in place and a large bandage applied. The hole will fill with spinal fluid. The bone will heal over in a few weeks and Neil will eventually become as hard-headed as ever.

Six months later, Neil was doing fine, having no seizures. About 60 percent of such epilepsy operations bring the seizures under much better control, although drugs often continue to be necessary, as in Neil's case.[1] He has already gone back to his engineering work, which had formerly been severely disrupted by his once-a-day seizures. Neil is looking forward to staying seizure-free for another six to twelve months, not so much because he fears an occasional seizure, but because he will then be able to get a driver's license again.

If all of the epileptics in the United States lived in one city, that city would be one of the largest in the country. But the same can be said for most of the disorders which we have mentioned in this book:

—2.5 million (1.2 percent of the population) are impaired from strokes
—2 million (1 percent) have epilepsy by one estimate, double that number by another estimate
—2 million have or will have schizophrenia during their lifetimes
—2 million have or will have a profound depressive disorder
—1.5 million (0.7 percent) have aphasia
—1 million (0.5 percent) have organic psychoses of toxic or neurologic origins and other permanently disabling mental disorders
—over a million elderly persons have senile dementia (of which at least 20-30 percent of the cases are preventable)
—another million suffer from disorders of the nerve-muscle synapse or muscle
—500,000 adults have Parkinson's disease
—200,000 have multiple sclerosis
—32,000 have tumors of the nervous system

This list does not include the 8.4 million (15 percent of school-age children) who suffer from reading or learning disorders, who have minimal brain damage, etc., or the 6 million mentally retarded persons in the United States. One could also add the estimated 50 million people in the United States who suffer from some form of chronic pain disorder (few, of course, as badly as Edith or the phantom-limb patient, but bad enough to lose a few days of work each year). One could conclude that a great many people have a personal interest in the progress of brain research.[2]

Understanding the basic principles on which the brain operates has, of course, a lot more to offer than just insight into the mechanisms through which diseases act. Like understanding the origins of the universe and the physical principles of energy and matter, understanding the brain is one of the great challenges of mankind. It is, so far as we know, the most highly organized three-pound bundle of matter in the universe.

Ambrose Bierce, in *The Devil's Dictionary*, gave the following definition:

> MIND, n. —A mysterious form of matter secreted by the brain. Its chief activity consists in the endeavor to ascertain its own nature, the futility of the attempt being due to the fact that it has nothing but itself to know itself with.

There are a lot of neuroscientists trying to prove him wrong. Also, many would dispense with the word "mind," and just stick to the word "brain." So far, it has been a distinction without a difference. The mind is merely a catch-all for the things that we can't explain otherwise, such as the common subdivision of pains into those that are "real" and those that are "all in your mind." The brain stem's internal morphine system for the regulation of pain sensation thus used to be part of the mind but it is now part of the brain.

Many people find it hard to begin thinking of the brain in computerlike terms. Some medical student can always be relied upon to ask a question (typically, "How does the upside-down image of the world, optically projected onto the back of the eye, ever get turned back right side up again?") which indicates an expectation that there is really a little person inside the head watching a TV screen. This "ultimate anthropomorphism" is a philosophical problem which tends to evaporate as one gets to know the details of the human brain better. With knowledge comes a great admiration and respect for the crowning achievement of several billion years of slow biological evolution, and a better insight into what has made possible the cultural phase of evolution, especially rapid in the five thousand years of written history. It is, of course, tempting to speculate that a detailed understanding of the brain would produce enhanced human capabilities, leading to yet another marked acceleration in the rate of evolution.

NOTES

General Reading List (suitable for most readers)

Asimov, Isaac. 1963. *The Human Brain, Its Capacities and Functions*. Hardcover, New York: Houghton Mifflin Company. Paperback, New York: New American Library, Mentor. Old but still quite useful.

Blakemore, Colin. 1977. *Mechanics of the Mind*. Cambridge: Cambridge University Press. A visual neurophysiologist, expert on developmental problems, prepared a BBC radio program series from which this book was adapted.

Bronowski, Jacob. 1973. *The Ascent of Man*. Boston: Little Brown. Written by a mathematician interested in form and function; adapted from the television series.

Gardner, Howard. 1976. *The Shattered Mind*. New York: Vintage. A book by a psychologist who is an expert on aphasia.

Harlow, Harry F. 1971. *Learning to Love*. New York: Ballantine Books. Developmental psychology; another version of "critical periods."

Miller, Jonathan. 1978. *The Body in Question*. New York: Random House. A general book on physiology, written by a physician turned entertainer.

Pines, Maya. 1973. *The Brain Changers, Scientists and the New Mind Control*. Hardcover: Harcourt Brace Jovanovich. Paperback: New American Library, Signet. A journalist's tour of research labs.

Thompson, Richard F., ed. 1972. *Physiological Psychology (Readings from Scientific American)*. San Francisco: Freeman. A good collection.

Wittrock, M. C., ed. 1977. *The Human Brain*. Englewood Cliffs, N.J.: Prentice-Hall. A collection of essays aimed at general readers but heavy on the terminology. Useful glossary.

Advanced Reading List (suitable for those willing to get technical)

Cooper, Jack R., Floyd E. Bloom, and Robert H. Roth.

1978. *The Biochemical Basis of Neuropharmacology.* 3rd ed., New York: Oxford University Press.

Eccles, John C. 1973. *The Understanding of the Brain.* New York: McGraw-Hill.

Hecaen, Henry, and Albert Martin. 1978. *Human Neuropsychology.* New York: Wiley.

Junge, Douglas. 1976. *Nerve and Muscle Excitation.* Sunderland, Mass: Sinauer.

Kandel, Eric R. 1976. *Cellular Basis of Behavior.* San Francisco: W. H. Freeman.

Katz, Bernard. 1966. *Nerve, Muscle, and Synapse.* New York: McGraw-Hill.

Kuffler, Stephen, and John G. Nicholls. 1976. *From Neuron to Brain.* Sunderland, Mass.: Sinauer.

Lund, Raymond D. 1978. *Development and Plasticity of the Brain, An Introduction.* New York: Oxford University Press.

McGeer, Patrick L., John C. Eccles, and Edith G. McGeer. 1978. *Molecular Neurobiology of the Mammalian Brain.* New York: Plenum Press.

Melzack, Ronald. 1973. *The Puzzle of Pain.* New York: Basic Books.

Penfield, W., and H. Jasper. 1954. *Epilepsy and the Functional Anatomy of the Human Brain.* Boston: Little, Brown.

Penfield, W., and L. Roberts. 1959. *Speech and Brain Mechanisms.* Princeton, N.J.: Princeton University Press.

Phillips, C. G., and R. Porter. 1978. *Corticospinal Neurons: Their Role in Movement.* New York: Academic Press.

Pincus, Jonathan H., and Gary J. Tucker. 1978. *Behavioral Neurology.* Second edition. New York: Oxford University Press. Seizure disorders, limbic system, schizophrenia, disorders of intellectual functioning, movement disorders, depression, sleep disorders, etc. An excellent book for professionals but also generally readable.

Schmidt, Robert F., ed. 1978. *Fundamentals of Neurophysiology.* Berlin: Springer Verlag.

Schmitt, Frank O., and Frederic G. Worden, ed. 1979. *The Neurosciences, Fourth Study Program.* Cambridge, Mass.: MIT Press. The earlier books in this series are also highly recommended.

Scientific American, special issue on the brain, September, 1979.

Shepherd, Gordon M. 1979. *The Synaptic Organization of the Brain*. Second edition. New York: Oxford Univ. Press.

Stevens, Charles F. 1966. *Neurophysiology: A Primer*. New York: Wiley. The last two chapters provide a transition into the quantitative aspects of membrane biophysics and are particularly recommended for readers with physical science backgrounds.

Williams, Moyra 1970. *Brain Damage and the Mind*. London: Penguin Books.

Chapter Notes

A name-year reference (e.g., Kandel 1976) is one which has been fully cited elsewhere, either in the reading lists or in another note.

Preface

The science-writer quotes may be found in: J. Schwartz, *Nature* 278:287, 1979. The author of the Latin-mass quote is James Gorman.

Chapter 1

1. A *transducer* is a device which converts one form of energy (such as pressure, heat, or light) into another, usually an electrical signal. "Sensor" is a somewhat equivalent word; in neurophysiology textbooks, the term "receptor" is typically used. We have used "transducer neuron" instead of "receptor neuron" not only because it is more precise but because the reader will encounter another use of the word "receptor" in a later chapter: the receptor molecules at the synapse, which combine with the neurotransmitter molecules to open the gates in the membrane of the next neuron.

2. Penfield and Jasper (1954) is the standard volume on localization of different functions in the human cerebral cortex, as well as the surgical treatment of epilepsy.

3. The distinction between motor and sensory cortex is relative rather than absolute: both areas show both types of responses, but motor responses are more frequent in front of the central sulcus (if present), sensory behind. For this reason, both areas are often collectively called the sensory-motor cortex. This cortex seems to be concerned

with fine, discrete rapid movements, such as those used in playing the piano. It has direct connections to nerve cells in the spinal cord that in turn directly activate muscles. There is another motor system that is involved with slower, grosser movements, those used in overall posture or control of neck and back movements. This includes an area of cortex on the inner side of the hemisphere known as the supplementary motor area (see Chapter 3), and multiple relays through clusters of nerve cells in the depths of the brain including striatum (See chapter 5) and cerebellum.

Animal studies suggest that the organization of sensory-motor cortex is quite complex. Sensation appears to come in from a particular site on the opposite side of the body in columns that extend through the full 2-3-mm thickness of cortex. This is predominantly touch and joint sensation, not pain. In monkeys, these columns seem to be arranged in several repeating face-arm-leg patterns extending in front of and behind the central sulcus (J. Kass et al., *Science* 204: 521-523, 4 May 1979). Under certain conditions, specific sites in this cortex are able to control the movement of single muscles (see Phillips and Porter 1978 or E. V. Evarts, *Scientific American* 229: 96, 1973). Although face and arm movements from cortical stimulation are almost always opposite to the side stimulated, the leg is another matter; in man there are also connections to the leg on the same side, as well as the opposite side. As a result, leg function commonly recovers better after damage to motor cortex than does arm or face.

4. An attempt has been made to use these flashes of light produced by stimulation of visual cortex as a way of conveying visual information to the blind, a "visual prosthesis." This was to be accomplished by placing many electrodes over visual cortex; each electrode, when stimulated, produces a small light-flash sensation. Each electrode is stimulated in a sequence determined by a television camera and computer. In this way, the blind person would "see" a sequence of flashes in the form of a letter or figure viewed by the camera. The technology to do this is available with activation of an array of stimulating electrodes through the intact skin by multiple very-short-range radio transmitters driven by a television camera. But in the cases where this has been attempted, the flashes that follow visual-cortex stimulation are in a different part of the visual field in each patient. As a result of this individual variability, a large external computer is needed to convert the image of a letter or figure to the right sequence of flashes for that patient, a requirement making the entire apparatus too clumsy for practical use at present. This individual variability not only involves function but also the anatomy of

the cortex; this variability has been carefully studied for visual cortex and turns out to be quite large. The total area of visual cortex varied by a factor of three, and the exposed surface fourfold, in twenty-five apparently normal brains (S. Stensaas, D. Eddington, and W. Dobelle, *Journal of Neurosurgery* 40: 747, 1974).

Nerve cells in visual cortex have a very detailed organization, discussed in Chapter 11.

The auditory system is no less complex. In the brain stem there are many interconnections between right and left ears, but closer to the cortex sounds from either side become separated, right-ear sounds going to left auditory cortex. A more successful neural-stimulation prosthesis has been devised for some of the deaf. It enters the auditory system not at the level of cortex but at the sound transducers in the inner ear. The spatial pattern of these nerve cells that transduce sound frequencies to nerve impulses is well known and fairly uniform from one patient to another. So nerve cells that would ordinarily send impulses to the brain when a particular sound frequency is heard can be artificially activated by electrical stimulation through this "cochlear prosthesis" when a microphone detects the same frequencies. Then a patient who is deaf from middle or inner ear damage, but whose nerve connections to the ear are intact, will be able to hear that sound.

Chapter 2

1. *Nerve* is a word used in many ways (and not merely in English: for example, the Hebrew word *etzev* means both "nerve" and "sadness"). In neurological terminology, a nerve is peripheral nerve such as the sciatic nerve, a large bundle of axons typically running from skin, muscles, etc. into the brain or spinal cord. The axons are sometimes called "nerve fibers"; a distinction can be made between the axon and the fiber (axon plus the myelin wrapping), but we shall use axon indiscriminately. When used as a modifier, "nerve" usually should be translated as "nervous": the "nerve impulse" is just the impulse, with no attempt being made to restrict it to the impulse of a peripheral axon.

2. *Impulse* has many synonyms: nerve impulse, spike, action potential, plus all of the verb forms such as when one speaks of the neuron "firing," "spiking," or "discharging." In this book, we have tried to just use "impulse" and "impulse production" or "triggering an impulse." The word "spike" is also used for another phenomenon seen in EEG recordings (a transient voltage fluctuation lasting tens of milliseconds, often seen in epileptic foci), but this is *not* synonymous with "impulse."

3.   Neurons usually work like one-way streets because the secretion process at a synapse is one-way: only the upstream neuron's axon terminals are specialized to release neurotransmitter, and only the downstream neuron has receptor molecules on its surface to detect the neurotransmitter. Thus, messages of the usual sort do not go in both directions. If impulses are initiated somehow in the axon terminals, an impulse is capable of spreading backward up the axon toward the cell body, but it is not known whether this capability is utilized by a typical neuron. Such backward spread is, however, seen in neurons whose axon terminals end in an epileptic focus; this "wrong way" travel may be one mechanism by which seizures spread (see Chapter 8).

4.   Evoked potentials are just the EEG at a particular time after a stimulus. Sometimes the evoked potentials are quite obvious, as in the alteration in the EEG after flashing a light in the eyes. Other changes are buried in the "noise" but, by repeating the stimulus dozens or hundreds of times, a signal can often be detected. Computers may be used to add together the EEG for a second following each light flash. This "averaging" will reveal some small voltage changes which occur 0.3 second after a stimulus. Indeed, this late wave behaves in a peculiar manner, depending upon whether the person was expecting the stimulus or not; for example, if each successive light flash alternates red-green-red-green, one can make the fourth late wave much bigger by throwing in an unexpected sequence such as red-green-red-red. A popularized treatment of the subject, complete with science-fiction extrapolations, can be found in R. M. Restak, *The Brain: The Last Frontier*, (New York: Doubleday, 1979).

Chapter 3

1.   *Strokes* are sudden, severe attacks; in the sense used in this book, they are events which damage a region of brain by "vascular accidents". There are numerous subclassifications. A "hemorrhage" is a leak which damages both by pressure and by depriving neurons of oxygen. A "thrombosis" is when the vessel is plugged up by a blood clot. An "embolus" is plugging by either a clot or other intravascular object (perhaps a plaque of atherosclerotic matter shed from a vessel wall somewhere upstream). An "aneurysm" is a weak spot on the wall of a blood vessel which bulges out like a balloon, occasionally killing nearby neurons by pressure (more often leaking to produce a hemorrhage). When things do not happen suddenly, they are not called strokes; tumors are one example, but so are

malformations of blood vessels which shunt arterial blood into veins before it perfuses the tissue. A "transient ischemic attack" (or TIA) is a stroke which causes no lasting (detectable) damage, usually because the vessel is unplugged before permanent damage is done; obviously the definition of a TIA is complicated by recovery of function. A "lesion" is an area of destruction, either natural or artificial in cause; for example, a surgeon is said to produce a thalamic "lesion" by heating during a thalamotomy for Parkinsonism. An "infarct" is a lesion caused by a vascular accident.

2. The example of a language deficit from a stroke in Wernicke's area, Blanche, is from H. Head, *Aphasia and Kindred Disorders of Speech*, v. 1, p. 305 (Reprint, New York: Hafner, 1963). The drawing of traditional language areas is adapted from T. Rassmussen and B. Milner, *Annals of N.Y. Academy of Sciences* 299: 355, 1977. Neil's individual naming and memory map follows Case 4, slightly modified by Case 3, of G. Ojemann and C. Mateer, *Science*, 205:1401–1403, 28 September 1979.

3. The first of our examples of a language disturbance after damage to Broca's area is from A. Luria, *Traumatic Aphasia* (The Hague: Mouton, 1970), p. 196. Broca's formulation not only included the first recognition of the role of the frontal lobe in language, but also the first public recognition that the left brain was especially important for language. Yet this formulation was based on remarkably shaky evidence. Indeed, the brain of his first case, Tan-tan, has only been looked at externally, and never cut. In that form it still resides in a Paris museum. So no one really knows the extent of the brain damage in that case. A century of subsequent observation generally was interpreted as supporting Broca, but recent studies suggest that the type of aphasia he described actually follows more widespread brain damage and not damage just to Broca's area (J. Mohr, *Studies in Neurolinguistics* 1: 201, 1976). A more modern concept of the organization of language cortex is presented in Chapter 4.

4. In practice *aphasia* is distinguished from "anarthria," aphasia meaning a disorder of language, anarthria a disorder in speech articulation from abnormal muscle movements of tongue, lips, face, and larynx. For example, language-cortex damage does not alter the muscle movements, it is just that the cortex sends signals for movements that make the wrong words—an aphasia. But cerebellar damage alters the coordination of the face and tongue muscles, so that speech articulation is jerky, but jerky on the right words—an anarthria.

5. *Conduction aphasia* is now thought to represent a failure in

short-term memory for words. See Chapters 4 and 6, and E. Warrington and T. Shallice, *Brain* 92: 885, 1969.

6. The location of brain damage disturbing color naming is from J. Mohr, J. Leicester, L. Stoddard, and R. Sidman, *Neurology* 21:1104, 1971.

7. Species-specific vocalization in the monkey also involves additional cortex on the inner side of the hemisphere, the cingulate gyrus (see Chapter 14) and part of the brain stem (Chapters 8-10; also see U. Jurgens, in H. Steklis and M. Raleigh, eds., *Neurobiology of Social Communication in Primates*, New York: Academic Press, pp. 11–44, 1979). Brain-stem structures may be important to emotional speech in man, but cingulate gyrus damage in man changes the general level of emotional responsiveness, rather than having any specific effect on emotional speech.

8. Individual patterns of naming are from G. Ojemann and H. Whitaker, *Brain and Language* 6: 239, 1978, and G. Ojemann, *Journal of Neurosurgery* 50: 164, 1979. The individual patterns of infoldings of cortex are also quite variable, especially in language areas. See, for example, A. Rubens, M. Mahowald, and J. Hutton, *Neurology* 26: 620, 1976. For ease of comparison, however, we have transferred all data to a standard brain sketch.

9. The various patterns of deficits in different languages after brain damage are reviewed in M. Paradis, *Studies in Neurolinguistics* 3: 65, 1977. Stimulation mapping in bilingual patients including the case illustrated is reported in G. Ojemann and H. Whitaker, *Archives of Neurology* 35: 409, 1978.

10. More detail on the patterns of reading deficits after brain damage may be found in H. Hecaen and H. Kremin, *Studies in Neurolinguistics* 2: 269, 1976. The interesting differential effect of language-area brain damage on reading the phonetic (Kana) and pictographic (Kanji) parts of Japanese is discussed in S. Sasanuma and O. Fujimura, *Cortex* 7: 1, 1971.

11. A *computerized tomographic scan* (or "CT scan") is an x-ray picture generated by a computer from many minutes of low-level x-rays taken from many different angles. The computer compiles a three-dimensional map of the brain based upon its density to x-rays; then, two-dimensional "slices" are calculated in the computer to show particular cross-sectional views. For example, a stroke will leave behind an area of lesser density due to the debris from the damaged brain; this can be spotted on a CT scan.

12. A right-brain role in recovery of language is documented in J. Pettit and D. Noll, *Brain and Language* 7: 191, 1979.

Chapter 4

In Chapters 4 and 15, we describe a number of research studies being carried out on Neil, one after the other. In practice, there isn't time available for all of them; a given patient might participate in only one or two of them. Neil's performance on the language-research studies described in this chapter is patterned after Case 4, modified somewhat to be more typical of findings in the overall series, by the findings in Case 3, of G. Ojemann and C. Mateer, *Science*, 205: 1401-1403, 28 September 1979. That paper is also the source of the data for the summary drawing. Note that Neil is a composite, and should not be cited as a case report.

1. C. Mateer and D. Kimura, *Brain and Language* 4: 262, 1977; and D. Kimura, *Studies in Neurolinguistics* 2: 145, 1976, discuss the changes in sequential oral movements, and in hand movements including use of sign language after damage to language areas of brain.

2. Psycholinguistic evidence that first led to the motor theory of speech perception is reviewed in A. Liberman, F. Cooper, D. Shankweiler, and M. Studdert-Kennedy, *Psychological Reviews* 74: 431, 1967. See also A. Liberman in *The Neurosciences: Third Study Program*, 1974, pp. 43-56.

3. E. DeRenzi and L. Vignolo, *Brain* 85:665, 1962.

Chapter 5

1. Some background material on Parkinson's-disease research can be found in: Conference Report #45 of the UCLA Brain Information Service, and *The Basal Ganglia* edited by Melvin D. Yahr (New York: Raven Press, 1976). The cell-loss data is on page 217.

2. Evidence for this specific language disturbance after left thalamic hemorrhage is to be found in: J. Mohr, W. Watters, and G. Duncan, *Brain and Language* 2: 3, 1975; G. Ojemann, *Studies in Neurolinguistics* 1: 103, 1976 (the source of the example given); A. Luria, *Brain and Language* 4: 432, 1977; and A. Reynolds, A. Harris, G. Ojemann, and P. Turner, *Journal of Neurosurgery* 48: 570, 1978.

3. This model of the effects of thalamic stimulation on language

and the evidence supporting it are to be found in G. Ojemann, *Brain and Language* 2: 101, 1975, and G. Ojemann, *Annals N.Y. Academy of Sciences* 299: 380, 1977.

4. *Levodopa* may have side effects such as psychosis or depression; see notes for Chapter 9. More common side effects are choreiform or dystonic movement disorders.

Chapter 6

Further reading on memory could start with L. R. Squire, "The anatomy of amnesia," *Trends in NeuroSciences* 3(3):52, March 1980; recent research collections include M. R. Rosenzweig and E. L. Bennett, editors, *Neural Mechanisms of Learning and Memory* (MIT Press: Cambridge), 1976, and M. A. B. Brazier, editor, *Brain Mechanisms in Memory and Learning* (Raven Press: New York), 1979. The limitations of parallel processing and the capacities of serial processing in human memory mechanisms are reviewed by H. A. Simon, *Models of Thought*, (New Haven: Yale), 1979, p. 4.

1. The finding that damage to both left and right hippocampus (and in surrounding brain on both sides) in man results in a severe memory loss was first reported in W. Scoville and B. Milner, *Journal of Neurology, Neurosurgery and Psychiatry* 10: 11, 1957. Evidence that this loss is specific to hippocampal damage is found in W. Penfield and G. Mathieson, *Archives of Neurology* 31: 145, 1974. The detailed studies on the effects of removal of one, or the other, hippocampus on short-term memory for different material are from Brenda Milner and her associates. A good review of those studies is in *The Neurosciences: Third Study Program*, 1974, pp. 75-89.

2. Why some patients show a very long period of retrograde amnesia, lasting years in H.M. and other patients with similar bilateral hippocampal damage, remains a puzzle. In animal studies, the transition from short-term to long-term memory, the process called consolidation, occurs in hours to several days, but is not thought to last years. No one has an explanation for the disappearance of memories from the several years before the hippocampal damage. Indeed, most researchers in the field ignore that finding.

3. *Maria* is adapted from Case 31 in W. Penfield and P. Perot, *Brain* 86: 595, 1963. This also is the source of the material on "experiential" responses with temporal-lobe stimulation and seizures.

4. The chemical changes that may be related to memory and the way chemicals influence memory are reviewed in: J. McGaugh in

vol. 1 of *Biological Foundations of Psychiatry*, ed. by R. Grenell and S. Gabay (New York: Raven Press, 1976).

5. *Anatomical changes.* See W. T. Greenough, "Development and memory: the synaptic connection," in *Brain and Learning*, ed. by T. Teyler (Stamford, Conn.: Greylock Publishers, 1978), pp. 127-145; and W. T. Greenough, "Experiential modification of the developing brain," *American Scientist* 63: 37-46, 1975.

6. *Pruning of spines.* See Lund et al. (1977) reference in notes for Chapter 11; also figure opening Chapter 12.

7. Selective conservation of random connections is just our variant of a decades-old notion called the "growth theory of learning" (see review in Chapter 15 of McGeer et al., 1978). Removal of nonparticipating synapses is particularly useful for explaining "one-trial learning," where only a single brief exposure to new information occurs, but a permanent memory develops. Multiple-trial learning, where the brain gets better and better at doing something with repeated exposures (as in motor skills), is often thought to be due to the strengthening of existing connections; the growth theory suggests synapses becoming bigger, stronger, etc. The latter is analogous to why the grass is greener over the septic tank; in the photographic analogy, one-trial exposure requires the image to be developed by removing unexposed elements, as in a drought leaving green grass only over the septic tank. Images from brief exposures can, of course, be held by other means (as in the image of a handprint on a frosted windowpane), but they need to be "fixed" somehow or they will be easily disrupted or overlaid. This has been the usual criticism made of the growth theory of synapses: neurons in the brain are active most of the time, and memory mechanisms such as synaptic facilitation would be expected to be both temporary and easily overlaid.

Another major issue is what constitutes synaptic "use." Is it just release of neurotransmitter by an impulse, or does it additionally require the next neuron to respond with an impulse (which could mean that it had other active inputs at the same time, c.f. the "conjunction theory of learning")? There have been proposals that some synapses regulate the plasticity of others (see end of Chapter 10) and that the hippocampus is involved in this regulation process.

Chapter 7

1. A good review of developmental dyslexia, including the genetic aspects, is A. Benton, *Advances in Neurology* 7: 1, 1975. The

dichhaptic test results in this condition are from S. Whitelson, in C. Shagass, S. Gershon, and A. Friedhoff, eds., *Psychopathology and Brain Dysfunction* (New York: Raven Press, 1977), pp. 15-50. Evidence of earlier and more complete lateralization of visual-spatial abilities in males can also be found in that paper. See also S. Whitelson, *Science* 193: 425, 1976.

2. The relation between handedness and language is reviewed in T. Rasmussen and B. Milner, *Annals of N.Y. Academy of Sciences* 299: 355: 1977.

3. It has long been known that left-handedness is much more common in those who stutter than in the overall population. Evidence of an unusually high incidence of bilateral language representation in stutterers, using the dichotic technique, has been reported by J. Brady and J. Benson, *Archives of General Psychiatry* 32: 1449, 1975. The cases in which damage to one side of the brain, in what would ordinarily be the language area on the left side, have cured lifelong stuttering are collected in R. Jones, *Journal of Neurology, Neurosurgery and Psychiatry* 29: 192, 1966.

4. That young brains have more flexibility is not surprising. It is not a matter of growing additional neurons but of the changes in size and complexity of existing neurons. The series of drawings which open each chapter tend to emphasize the changes in the dendritic tree during early childhood, and the branchings of the axon are likewise elaborated during development, as shown in the drawing opening Chapter 13. The myelin covering of many axons is not complete until late childhood.

Chapter 11 discusses critical periods in development in which visual experience in infancy seems to make permanent wiring changes in the visual cortex. There is a great deal of research on how the brain gets wired up during prenatal development: axons grow from the cell body toward a target, but how does the axon know where to go? We discuss this tissue tangentially in Chapter 13 when we cover regeneration of damaged axons in adults. For a modern treatment of developmental plasticity, see R. D. Lund, *Development and Plasticity of the Brain* (New York: Oxford University Press, 1978).

5. The grammatical deficits in the speech of children whose left hemisphere had been removed in infancy are from the report of M. Dennis and H. Whitaker, *Brain and Language* 3:404, 1976.

6. The greater asymmetry in the size of the planum temporale in males than in females is only one example of sex differences in brain

structure. Sex hormones clearly enter the brain (at least in animal studies) and have specific receptor sites in several brain areas, including hypothalamus and amygdala (see Chapter 14). Animals have a period in early life when exposures to sex hormones irreversibly alter the animal's sexual behavior. For example, brief exposure of a female rat to male sex hormones in the first few days after birth leads to a loss of ovulation and to the failure of female mating behavior in adulthood. These animals show the male pattern in the number of neurons in the preoptic area of the hypothalamus, rather than the expected female pattern. So this brief exposure to the opposite sex hormones during a "critical period" of infancy (see also Chapter 11) has changed both brain anatomy and sexually related behavior. It is not clear whether humans have such critical periods for sex hormone effects; if so, they probably occur before birth. However, there is some evidence that human sexual identity depends upon sex hormone levels, rather than being socially determined. There is a rare hereditary condition where an enzyme is partially missing that allows the male sex hormone to stimulate the formation of the male external genitalia. Children with this disorder look like females when born, but have male sex hormone levels in their blood. Because their external genitalia have the female form, they are raised as girls. But at puberty, these children make enough male sex hormone so that their external genitalia become more like males. Though raised as girls up to this time, nearly all of these children now act like, and identify themselves as males. It has been suggested that the exposure to the male sex hormone before birth or in childhood determines the ultimate sexual identity, regardless of cultural influences. In another condition, congenital adrenal hyperplasia, an excess of male sex hormone is produced. Female humans (and monkeys) with this condition act in a highly masculine way, e.g., spending little time at maternal role rehearsal, though these girls identify themselves as female. See *Science* 205:985, 1979 for a review of these studies; and J. Imperato-McGinley et al, *New England Journal of Medicine* 300:1233, 1979 (and the accompanying editorial by Wilson in the same issue, p. 1269).

7. The anatomical asymmetries between left and right cerebral cortex have been the subject of many research reports. Of particular importance are those of N. Geschwind and W. Levitsky, *Science* 161: 186, 1968; and J. Wada, R. Clarke, and A. Hamm, *Archives of Neurology* 32: 239, 1975. See N. Geschwind, "Specializations of the human brain," *Scientific American* 241 (3):180-199, September, 1979.

8. A discussion of Woodrow Wilson's later years may be found in

G. Smith, *When the Cheering Stopped*, New York: William Morrow and Co., 1964. A comprehensive review of the deficits associated with damage to the nondominant hemisphere can be found in R. Joynt and M. Goldstein, *Advances in Neurology* 7: 147, 1975.

9.   A particularly good discussion of the effects of brain damage on artistic abilities is found in Chapter 8 of H. Gardner, *The Shattered Mind* (New York: Vintage, 1976). This is also the source for some of the information on lateralization of musical abilities. The dichhaptic test lateralization of mathematical skills is from S. Diamond and J. Beaumont, *Psychonomic Science* 26: 137, 1972. A more general review of the brain basis for mathematical calculations is found in F. Grewel, in Vol. 4 of *Handbook of Neurology*, ed. by Vinken and Bruyn (Amsterdam: North Holland, 1969), pp. 181-194. The paper by N. Wertheim in the same volume (pp. 195-206) reviews additional literature on music defects after brain injury. The effects of intracarotid amobarbital on music are reported by H. Gordon and J. Bogen, *Journal of Neurology, Neurosurgery and Psychiatry* 37: 727, 1974.

10.   Blood-flow changes during different activities are reviewed in N. Lassen, D. Ingvar and E. Skinhoj, *Scientific American* 239 (4): 62, 1978. Our figure was redrawn in black and white from one of their elegant color computer displays showing blood-flow changes during silent reading.

Chapter 8

1.   Newspaper reports of violent behavior by epileptics are often gross oversimplifications. As we discuss in Chapter 9, some temporal lobe epileptics also have schizophrenia. Just because someone is taking anticonvulsant medications or has an abnormal EEG, and also has psychotic behavior, does not mean that the epilepsy caused the behavior. Curiously, schizophrenics with temporal lobe epilepsy may experience even more severe psychotic symptoms when the seizures are brought under control by anticonvulsants. Sometimes, seizures make depressive and schizophrenic symptoms better; see the rationale for electroconvulsive shock treatments in Chapter 14.

2.   *Local sign*. Epileptic foci in the frontal lobe produce no aurae or other local signs. To get an impression of the difficulty in detecting an aura, imagine a focus in a "reading" area of the parietal lobe (Chapter 3): the patient would be able to detect the forewarning of the seizure only if reading at the time.

3.   There is a chance that someone like Patty could never make up

the lost educational opportunities. One cannot always make up for lost time. There are skills which are not learned well unless learned at the right age, as a later chapter shows. Bureaucratic delays in the release of an effective anticonvulsant drug might prevent someone from achieving seizure control during such a "critical period" for an essential skill. This is the other side of the coin for the "prohibit it until absolutely sure it's safe" drug legislation and administrative practices in the United States. While these regulations protect against such calamities as another thalidomide tragedy, they can also lead to a series of tragedies related to witholding effective therapies.

The FDA has tended to be slow to approve new drugs for some good reasons: the laws established by Congress make it very difficult for the FDA to withdraw a drug after approval or to limit its use (e.g., only by non-pregnant persons, only prescribed by specialist physicians in research centers, etc.). Another serious impediment is the great expense of testing a drug with a limited market, e.g., only for the 14,000 persons in the U.S. suffering from Huntington's chorea.

Chapter 9

Our brief exposition on synaptic transmission may be supplemented by consulting the advanced reading list, especially Katz (1966), Kandel (1976), and Kuffler and Nicholls (1976). They do not cover one important aspect of synaptic transmission, especially important for those neurons which lack impulses (see notes on Chapter 11). Axon terminals may continuously leak neurotransmitter at a rate proportional to the local voltage in the axon terminal. While impulses invading the terminal are one source of voltage, providing a brief squirt of neurotransmitter, voltages from nearby synapses onto the axon terminal (as in the "third neuron" figure in Chapter 10) also slowly modulate the leakage rate. See K. Graubard and W. H. Calvin, "Presynaptic dendrites: implications of spikeless synaptic transmission and dendritic geometry", Chapter 18 in: *The Neurosciences, Fourth Study Program*, ed. by F. O. Schmitt and F. G. Worden (Cambridge, Mass. MIT Press, 1979). In Chapter 29 of that book, entitled "Styles of neuronal computation," the same authors discuss the neuron as a computing machine.

1. For more on sleep, see the book by one of the leading sleep researchers: W. C. Dement, *Some Must Watch While Some Must Sleep* (San Francisco Book Co., 1976).

2. W. Berlin (personal communication) has noted a minor motor-movement disorder called chorea in about a third of children labeled

as minimally learning-disabled. Chorea is thought to result from malfunctions of the striatum.

3.   Bleuler, quoted in Crow (1979; see below).

4.   One of the predictable problems with levodopa therapy for Parkinson's disease is that it could produce too much dopamine in other neurons which don't need it, i.e., forced feeding. About 15 percent of patients treated with levodopa suffer psychoses as a side effect, also suggesting a role of dopamine in schizophrenia.

5.   *Dopamine.* A number of different antischizophrenic drugs block receptors at dopamine synapses. Indeed, one of their side effects, as might be expected, is to temporarily induce some of the symptoms of Parkinson's disease. Dopamine may be only part of the problem in schizophrenia; it is thought to contribute to the stereotyped behavior. An increase in dopamine effectiveness does not, however, appear to account for other prominent symptoms, such as withdrawal and autistic behavior, flatness of affect, and lack of motivation. These symptoms suggest that another transmitter, norepinephrine, may be involved. Norepinephrine is made from dopamine with the aid of an enzyme; some research groups think that there may be a deficiency in the enzyme, thus causing an imbalance in the relative amounts of norepinephrine and dopamine. See S. S. Kety, "The biological substrates of abnormal mental states," *Federation Proceedings* 37: 2267-2270, July 1978.

6.   The *receptor excess* is discussed by T. J. Crow, "What is wrong with dopaminergic transmission in schizophrenia?", *Trends in Neurosciences* 2(2): 52-55, February 1979. For some of the heated arguments about whether dopamine and norepinephrine are elevated in schizophrenics, see E. D. Bird et al, *Science* 204: 93-94, April 1979.

7.   A good example of the *wrong key* plugging the keyhole is the paralyzing drug curare. It latches onto the receptors for the neurotransmitter acetylcholine, thus preventing acetylcholine released from axon terminals from combining with the receptor to open the gates for ion flows through the membrane. Acetylcholine is used at the nerve-muscle synapse, hence the paralyzing action of curare. The *nerve-muscle synapse* is often called the neuromuscular junction. LSD is thought to block receptors for serotonin. The antischizophrenic drug thorazine blocks dopamine receptors; Benadryl and Dramamine block histamine receptors. See A. Nathanson and P. Greengard, "Second Messengers in the Brain," *Scientific American*, August 1977, pp. 108-119.

8. *Slowed cleanup.* One way in which the cleanup works is to reabsorb the neurotransmitter back into the axon terminal for recycling. The tricyclic antidepressants such as Elavil slow that reabsorption, thus allowing neurotransmitters such as norepinephrine and serotonin to work longer upon the next neuron following an impulse.

Chapter 10

1. *Analgesics turning down sensitivity.* See R. H. Gracely, R. Dubner, and P. A. McGrath, "Narcotic analgesia: Fentanyl reduces the intensity but not the unpleasantness of painful tooth pulp sensations," *Science* 203: 1261-63, 23 March 1979.

2. *Endorphins and placebo responders.* See J. B. Levine, N. C. Gordon, R. T. Jones, and H. L. Fields, "The narcotic antagonist naloxone enhances clinical pain," *Nature* 272: 826-827, 27 April 1978. There is a similar story for acupuncture: see D. J. Mayer, D. D. Price, and A. Raffil, "Antagonism of acupuncture analgesia in man by the narcotic antagonist naloxone," *Brain Research* 121: 360-373, 1977. One Seattle newspaper headline read, "Reign of pain falls mainly in the brain."

3. *Depression* is a word used in quite different ways by psychiatrists and synaptic physiologists. No causal relationship is to be inferred.

4. *Aplysia.* The presentation of the habituation/dishabituation story here is quite simplified. For an accessible exposition by one of the leading researchers, see Kandel (1976).

5. *Regulation of plasticity.* Regulation of synaptic transmission has been studied in the habituation of various escape reflexes, such as those of crayfish and lobster tail flip. See F. B. Krasne, "Extrinsic control of intrinsic neuronal plasticity: a hypothesis from work on simple systems," *Brain Research* 140: 190, 1978.

Chapter 11

1. The neuron chain for visual, somatic, taste, smell, and auditory sensations may be found in any number of textbooks of anatomy or physiology; for a picture-book version, try F. H. Netter, *Nervous System*, Vol. 1 of *CIBA Collection of Medical Illustrations*, CIBA 1962.

2. Our presentation of the retina is quite abbreviated. There are many neuron types other than those illustrated (see Kuffler and Nicholls 1976). Except for the retinal ganglion cells and the amacrine

cells (the octopus-shaped cell on page 115), all of the neurons of the retina manage to function without producing impulses: they release neurotransmitter continuously and vary the leakage rate with light intensity or net synaptic input voltages. These neurons are not elongated enough to require booster stations; all neurons with long myelinated axons use impulses. See notes on page 197.

3. *Surrounding ring.* This doughnut-shaped map is called a "center-surround receptive field" in the literature. Some cells have excitatory centers and inhibitory surrounds, such as the example given in the text and the figure. Other cells have an excitatory surround and an inhibitory center, so that the optimal stimulus is a black spot on a light background, appropriately positioned. Similarly, many of the cortical neurons prefer black lines on light backgrounds. The size of a third order neuron's receptive field in monkey retina is about 1°, twice the arc subtended by the sun or moon (see Hubel and Wiesel 1977). On a monkey retina, this is about 0.3 mm. While the surround's diameter is usually about 1°, the center size varies greatly. For cells in the fovea (the high resolution center of the visual field which we use to read), the center may be only a few minutes of arc. For further reading, try D. H. Hubel and T. N. Wiesel, "Brain mechanisms of vision," *Scientific American*, 241 (3):150-162, September, 1979; and Kuffler and Nicholls (1976).

4. *Fourth neuron.* Neuroscientists do not actually use the "Nth neuron" notation to identify the position of the neuron in the chain. This is because it is not really a simple chain. In the retina, the retinal ganglion cells which send their axons to the thalamus (which we have called third-order neurons) can also be considered fourth or fifth order, as there are some circuitous pathways to them besides the direct path. One often hears the phrase "every cell is connected to every other cell—either directly or indirectly." Details of intracortical circuitry in monkeys can be found in Hubel and Wiesel (1977); for cats, in C. D. Gilbert and T. N. Wiesel, "Morphology and intracortical projections of functionally characterised neurones in the cat visual cortex," *Nature* 280:120-125, 1979.

5. *Fair-good-excellent responses.* To help the reader, we have omitted the time aspect: the responses we illustrate are merely those which occur when the light spot or bar is first turned on. Many neurons respond equally well when the light is turned off, usually best from an area adjacent to where the light-on response is optimal. For example, inhibitory surrounds are inhibitory during the light; when the light goes off, a few impulses occur afterward. Moving a light from such a surround into the center is often the very best way

to get a big response, because the light-off response occurs at the same time as the light-on response. Moving stimuli are usually more effective than stationary stimuli, and some cortical neurons will respond only to moving stimuli. Cortical neurons are often particular about the direction of movement, e.g., giving a better response to left-to-right movement than right-to-left. To understand this better, the reader must learn to work with the concept of receptive fields (see earlier note), which we have avoided here because students are apt to be initially confused by it. See Kuffler and Nicholls (1976).

6. The center-surround receptive fields in cerebral cortex come from cells in layer IVc, the layer in which most of the incoming axons from the thalamus terminate. The simple line-oriented receptive fields are found in cells of layer IVb (in monkey; in cat, they are also found in layer VI). The complex and hypercomplex receptive field types in monkey come from cells in layers II, III, V, and VI. About half of the complex and hypercomplex cells respond to both eyes; in contrast, almost none of the center-surround cells or simple cells are "binocular". See Hubel and Wiesel (1977, 1979).

7. *Rangefinders* work by measuring the tilt in a mirror which is needed to bring the two images into perfect overlap. In most cameras, as with the human eye and brain, it is hard to tell the difference between five meters (about sixteen feet) and infinity, as the two correspond to only a small movement of the adjustment knob. In the case of the eyes, the amount of muscle effort needed to produce convergence similarly gives a clue about the distance of objects. Another clue is the degree of disparity when there isn't perfect overlap. Some of the neurons in the visual association cortex seem specialized for detecting such differences in the images from the two eyes; see S. M. Zeki, "Cells responding to changing image size and disparity in the cortex of the rhesus monkey," *Journal of Physiology* 242: 827-841, 1974.

The figures show the optimal position of the light at the +, but + is not the fixation point. Each part of the visual cortex represents a different point in the visual field. The + is that point for the neuron under study.

8. *Deprivation studies.* Several simplifications have been made in our treatment of the monocular deprivation studies to aid comprehension. A frosted contact lens has not been used in most experiments, but rather sewing the eyelid shut; both procedures allow diffuse light into the eye but prevent form perception. Rather than alternating occlusion with contact lenses, some of the relevant experiments have instead been done with animals made artificially cross-eyed; again,

both eyes participate in visual experience but the images are seldom in register so that binocular coordination is difficult. See Hubel and Wiesel (1977; see below); and C. Blakemore, "The conditions required for the maintenance of binocularity in the kitten's visual cortex," *Journal of Physiology* 261: 423-444, 1976.

9. *Human critical period.* The reader may be curious as to how critical-period information can be obtained on binocularity in humans. There is a visual illusion called the tilt aftereffect: if a subject stares for several minutes at some tilted lines with one eye, it will cause vertical lines seen subsequently with the other eye to tilt too. In people with abnormal early visual experience due to squint but in whom the cross-eyed condition was surgically repaired, the tilt transfer aftereffect is much less noticeable. The earlier the surgical repair, the more normal the aftereffect. The binocular transfer aftereffect is most normal at later ages if the children had been operated upon in the first two years of life. In primates, the image from the two eyes is not compared until the visual cortex (at what we have called the sixth-order neuron), so the illusion must be generated there or subsequently. See M. S. Banks, R. N. Aslin, and R. D. Letson, "Sensitive period for the development of human binocular vision," *Science* 190: 675-677, 1975.

10. The evidence for a critical period for language comes from the study of children raised in severely deprived environments, especially a girl named Genie who was not regularly exposed to speech until age thirteen. Her language has not developed normally, and that which has is not normally lateralized, suggesting that the critical period for language acquisition passed her by. See V. Fromkin, S. Krashan, S. Curtiss, D. Rigler, and M. Rigler, *Brain and Language* 1: 81, 1974.

11. D. H. Hubel, "Effects of distortion of sensory input on the visual system of kittens," *The Physiologist* (American Physiological Society) 10: 43, 1967.

12. The analysis of the monkey's primary visual cortex (area 17) suggests that it is comprised of hundreds of elementary building blocks of cortical circuitry, which are themselves no larger than 1-2mm². Each such block contains the neural machinery to analyse all possible angles, from both eyes, for the small region of the visual field which projects to it (as much as 3° of arc if the block represents a region of the peripheral field, only 0.1° if the fovea). As Hubel and Wiesel (1977) note: "Visual perception, then, can in no sense be said to be enshired in area 17—the apparatus is simply not made to

analyse a percept that occupies more than a small region of visual field. All of the single cell physiology in fact suggests that area 17 is concerned simply with what may be thought of as the building blocks for perception."

The development work is reviewed in the Ferrier Lecture of D. H. Hubel and T. H. Wiesel, "Functional architecture of macaque monkey visual cortex," *Proceedings of the Royal Society, London, Series B*, 198: 1-59, 1977. For the anatomy of the visual cortex, see J. S. Lund, "Organization of neurons in the visual cortex, area 17, of the monkey (Macaca mulatta)," *Journal of Comparative Neurology* 147: 455-496, 1973; and J. S. Lund, R. G. Boothe, and R. D. Lund, "Development of neurons in the visual cortex (area 17) of the monkey (Macaca nemestrina): A Golgi study from fetal day 127 to postnatal maturity," *Journal of Comparative Neurology* 176: 144-188, 1977. The latter paper demonstrates the "budding" of spines during infancy, as discussed in Chapter 6 (see figure beginning Chapter 12).

Chapter 12

Because it is simpler to teach, we have illustrated impulse propagation with saltatory conduction (from the Latin *saltare*, to leap), where the impulse skips over the myelin-insulated regions to excite only the nodes. Many small axons do not have myelin insulation; they conduct impulses in a continuous wavelike manner analogous to a slowly burning fuse. When the myelin is lost from a normally myelinated axon, as in multiple sclerosis, conduction usually halts, as the gap is too large to leap over. However, some demyelinated regions have now been shown to develop enough voltage-sensitive channels in the denuded area to support the fuselike continuous conduction (H. Bostock and T. A. Sears, *Journal of Physiology* 280:273, 1978).

1. The diagnosis of multiple sclerosis (MS), without an indication such as tic douloureux in a young adult, is often difficult. Evoked potentials (EEGs following a light flash or other sudden stimulus) have proved quite helpful. The slowed speed of the impulse going down an axon means that the visual cortex gets the light-flash message later than usual. This delays the evoked potential peaks. Such delayed peaks help to diagnose MS more reliably than most other tests. Relatively minor amounts of demyelination only slow down the impulse; slightly more extensive patches of demyelination may block impulses altogether, leading to blind spots in the visual fields.

2. The *impulse* rises about 100 millivolts. It starts, however, from either 0 or from −70 millivolts, depending upon how one references

the voltage. We have simply referenced it to the internal voltage of the neuron at "rest," before an impulse. More commonly, it is referenced to the voltage outside the neuron. The internal voltage is then -70 millivolts at rest, rising to $+30$ during an impulse. At rest, the membrane is much more permeable to potassium than to sodium. If the membrane were permeable only to potassium, the internal voltage would be -90 millivolts; were the membrane permeable only to sodium, it would be about $+60$ millivolts. These "batteries" are created by the ion-concentration differences. Thus -70 is the compromise dictated by a 7:1 permeability ratio. Actually, things are somewhat more complicated, since chloride and calcium must also be considered.

3.   Much more could be said about how neurons control their impulse production rates and use pattern codes. For references, see W. H. Calvin, "Setting the pace and pattern of discharge: do CNS neurons vary their sensitivity to external inputs via their repetitive firing processes?" *Federation Proceedings* 37: 2165-2170, June 1978; and W. H. Calvin and D. K. Hartline, "Retrograde invasion of lobster stretch receptor somata in control of firing rate and extra spike patterning," *Journal of Neurophysiology* 40: 106-118, January 1977.

4.   The specific patterns of nerve-cell discharges, said to be associated with specific words or groups of words in a given individual, have been recorded from thalamus and reported by N. Bechtereva, P. Bundzen, Y. Gogolitsin, V. Malyshev, and P. Perepelkin (*Brain and Language* 7: 145, 1979) of the Institute of Experimental Medicine, Leningrad.

5.   *Charge comes from nearby naked membrane.* The reader with a working knowledge of physics may recognize this circumlocution: the impulse must charge up the capacitance of the 1 mm of membrane between nodes. The myelin wrapping performs the series capacitor trick, serving to reduce capacitance. This will reduce the charge wasted on capacity and increase the charge contributing to voltage alteration at the next node. Demyelination, by removing the wrapping, means that much charge must be wasted upon charging up the membrane capacitance.

6.   *Cooling nerves.* The longer the impulse lasts, the more time that there is to attract charges away from the next node (capacitance, again). See C. L. Schauf and F. A. Davis, "Impulse conduction in multiple sclerosis: a theoretical basis for modification by temperature and pharmacological agents," *Journal of Neurology, Neurosurgery and Psychiatry* 37: 152-161, 1974.

7. The *calcium effect* is complex, involving a layer of surface charge on the membrane. It misleads the voltage-sensing mechanisms inside the membranes as to the true voltage between the interior and exterior of the cell. Some drugs, such as aspirin, are thought to modify this surface layer and thus the "bias" upon the neuron.

8. For more on membranes, see Kuffler and Nicholls (1976), Katz (1966), Kandel (1976), or C. F. Stevens, "The neuron", *Scientific American* 241(3):55-65, September, 1979.

Chapter 13

1. *How caterpillars turn into butterflies* happens to be extremely relevant to recovery of function after strokes. Imagine yourself a neurobiologist who wanted to discover how a neuron could completely alter the function which it subserves: what better way than to study the neurons which make a caterpillar crawl. Those neurons lose their connections to the crawling system and make a whole new set of connections to other neurons so that they can run the butterfly's wings. It turns out (see J. W. Truman and S. E. Reiss, *Science* 192: 477, 1976) that hormones control certain steps in this conversion; there are several million people with strokes in the U.S. who might reasonably wish that a doctor could give them a hormone which would allow neurons in undamaged regions of their brains to alter, so as to pick up the functions lost in the stroke. Few scientists have any difficulty in seeing the potential connections between such pure biology and such therapeutic possibilities. Yet, every year a politician will hold up "caterpillars turning into butterflies" types of research as a prime example of how the government wastes money. Every year, the newspapers continue to print such stories uncritically, as they make a good headline on a slow news day and help to sell newspapers. Science should be no more immune to fair criticism than any other endeavor; typically, however, neither the critic nor the newspaper will show any interest in what the scientist's motivations were, or in what other informed people might think of the research.

2. The competition between sprouting and regrowing nerves can be found in: M. Devor, D. Schonfeld, Z. Seltzer, and P. D. Wall, "Two modes of cutaneous reinnervation following peripheral nerve injury," *Journal of Comparative Neurology* 185:211-220, 1979. Similar sprouting, and the suppression of it by healthy neighboring neurons, is thought to occur in the brain. While the "weathered" neurons seen in Chapter 6 are perhaps the most striking features of aged

brains, a careful analysis shows that many of the remaining neurons have *longer* dendritic trees than those in middle-aged brains, as if those neurons had expanded into the territory of shrunken neurons. Indeed, in the brains of patients dying with senile dementia, the sprouting mechanism seems to have failed (see S. Buell and P. Coleman, *Science* 206:854-856, 1979).

3. Here we speak of impulses being "canceled" during their journey along the axon, as it brings to mind canceling an order: neurophysiologists talk instead of the "intermittent conduction" of the impulse (see the article by I. Parnas in *The Neurosciences, Fourth Study Program* in the advanced reading list).

4. Our discussion of discs pressing upon spinal nerves to cause leg and foot pain does not make an important distinction. The region of the spinal nerve which passes near the disc is a region which is an exception to the general rule: it is pressure-sensitive even in normal nerves (see figure on page 148). Nerves damaged or irritated for some days can acquire the ability to initiate impulses in additional regions, as discussed in the text. Both the normal and the abnormal mechanosensitivity are discussed in J. F. Howe, J. D. Loeser, and W. H. Calvin, "Mechanosensitivity of dorsal root ganglia and chronically injured axons: a physiological basis for the radicular pain of nerve root compression," *Pain* 3: 25-41, February 1977.

5. For more on neuralgias, including tic douloureux, there are no elementary references. The technically inclined may find references in: W. H. Calvin, "Some design features of axons and how neuralgias may defeat them," in *Advances in Pain Research and Therapy*, Vol. 3, (New York: Raven Press, 1979). An excellent research collection is *Physiology and Pathobiology of Axons*, ed. by S. G. Waxman (New York: Raven Press, 1978).

Chapter 14

1. *Wisconsin card-sorting task*. Because of its familiarity to the average reader, we have used an ordinary deck of playing cards to illustrate this card-sorting task. In practice, neuropsychologists use a special deck of cards which allows a wider variety of sorting strategies. The effect of frontal-lobe damage on this and similar tasks has been reported by Brenda Milner in *Frontal Granular Cortex and Behavior*, ed. by J. Warren and K. Akert (New York: McGraw-Hill, 1964).

2. The limbic system was first defined anatomically by James Papez (*Archives of Neurology and Psychiatry* 38: 725, 1937) to include structures on the inner side of temporal lobe: amygdala,

hippocampus, and hippocampal gyrus; structures on the inner side of the frontal lobe; cingulate gyrus; and the connections between these through hypothalamus (mamillary bodies) and thalamus (anterior nucleus). This limbic lobe forms a ring of cortex around the upper end of the brain stem, and contains cortex whose structure suggests that it arose earlier in evolution than the other parts of cortex. Evidence that the sites of human brain damage that alter emotion are in the limbic lobe is reviewed by K. Poeck in vol. 3 of *Handbook of Neurology* (Amsterdam: North Holland, 1969), pp. 343-367.

3. D. Bear and P. Fedio, *Archives of Neurology* 34: 454, 1977.

4. A considerable literature is available on effects of brain damage on recognition of faces. See especially B. Milner, *Neuropsychologia* 6: 191, 1968. The relation of lateralization to one's perception of others' faces is discussed in C. Gilbert and P. Baker, *Neuropsychologia* 11: 355, 1973. Their study indicates that it is the right half face that we identify as *that person's face*, not the left half. (A person's right half face falls in our left visual field, related to our right brain). Lateralization of laughing and smiling to right brain is reported by F. King and D. Kimura, *Canadian J. Psychology* 26: 111, 1972. Evidence that emotions are likely to be perceived in the left visual field, and expressed in left face, is summarized in R. Campbell, *Cortex* 14: 327, 1978.

5. The evidence for a genetic basis for the major psychoses and possible biochemical abnormalities in these conditions is reviewed in Pincus and Tucker, 1978.

6. Evidence for the association between temporal-lobe epilepsy and psychoses is presented in S. Waxman and N. Geschwind, *Archives of General Psychiatry* 32: 1580, 1975. The abnormal collections of gray matter in patients with both temporal-lobe epilepsy and schizophrenia, called "hamartomas," are described in M. Falconer, *New England J. Medicine* 289: 451, 1973.

7. The biochemical basis of psychoses was reviewed briefly in notes for Chapter 9; see Kety reference.

8. In the mid-1970s, the critics of psychosurgery helped persuade the U.S. Congress to establish a National Commission for the Protection of Human Subjects of Biomedical and Behavioral Research and to charge it, among other things, with evaluating psychosurgery; see news article "Psychosurgery: National Commission issues surprisingly favorable report," in *Science* 194:299, 15 October 1976.

Both its report and the response to the report by Health, Education and Welfare Secretary Joseph Califano were generally favorable to psychosurgery; Secretary Califano concluded that these operations are useful in selected cases, provided that the patient is competent to give informed consent, and that regulation should be left to the medical profession, aided by voluntary standards set by a voluntary professional advisory body. See *Federal Register* 43(221): 53242, 1978. For an article on the political controversy surrounding this treatment, see Henry Miller, *New Scientist* 55: 188, 1972.

A review of the extensive literature on psychosurgery, along with a detailed account of the events leading up to the first "prefrontal lobotomy" (the preferred term is "leucotomy," which means cutting the bundles of axons) may be found in: E. Valenstein, *Brain Control*, (New York: Wiley, 1973). Valenstein also provided an updated review with 629 references for the National Commission report (*Psychosurgery*, appendix, DHEW Publication OS77-0002).

Valenstein estimated that psychosurgical operations are performed two to three times more often, per capita, in Great Britain and Australia than in the U.S. Among the British studies of the modern frontal-lobe operations are: Tan, Marks, and Marset, *British Journal of Psychiatry* 118: 155, 1971; Marks, Birley, and Gelder, *British Journal of Psychiatry* 112: 757, 1966. In these studies, detailed followups were done five years after operations for depression, obsession, and phobias; these patients' followups were compared with those of nonoperated patients with similar illness (although controls who were as sick as those coming to operation could not always be found). A retrospective study in a large series is Strom-Olsen and Carlisle, *British Journal of Psychiatry* 118: 141, 1971. For the studies where objective behavioral and intelligence tests were used before and after operations, see Mitchell-Heggs, Kelly, Richardson, and McLeish, pages 327-336 in *Modern Concepts in Psychiatric Surgery*, ed. by Hitchcock, Ballantine, and Meyerson (Amsterdam: Elsevier, 1979); Mitchell-Heggs, Kelly, and Richardson, *British Journal of Psychiatry* 128: 226, 1976; and Corkin, Twitchell, and Sullivan, pages 253-272 in *Modern Concepts in Psychiatric Surgery*.

All these reports and our discussion deal with operations in the frontal lobe for depression, obsessions, and phobia. Episodic rage is another condition for which psychosurgical operations have been attempted. Few people have ever witnessed such rage behavior, as it is not the anger which is likely to be encountered in barroom brawls or street demonstrations. These rare patients may be triggered into trying to kill a stranger by a stimulus as innocuous as a touch on the coat sleeve. Such patients may be too dangerous to be allowed near other people, even in an institutional environment. There is some evidence that such behavior follows damage to the amygdala (which is near the uncus, on the diagram in Chapter 14), and that removal of this damaged area will reverse this episodic rage behavior (while not preventing normal anger). Evaluations of amygdala surgery for this problem can be found in Hitchcock and Cairns, *Postgraduate Medicine* 49: 894, 1973; also Mark, Erwin, and Sweet, pages 379-391 in *Neural Basis of Violence and Aggression*, ed. by Fields and Sweet (St. Louis: Warren Green, 1975).

It has also been suggested that the cerebellum may be involved in psychoses. Unlike the other sites mentioned here, the cerebellum is not in the limbic system; it is usually thought to be a coordination center for motor control. However, removal of the midportion of the cerebellum prevents the development of abnormal rage behavior in one of the animal models of the illness, that of monkeys raised without either real or surrogate mothers. See the discussion of the studies of A. J. Berman, D. Berman, and J. W. Prescott in: R. Dow, *Mount Sinai Journal of Medicine* 41: 103, 1974. It has been suggested that electrical stimulation of the surface of the cerebellum improves some psychoses; whether this therapy is really effective, and if so for what types of behavioral problems, remains to be determined. See Heath, Llewellyn, and Rouchell, pages 77-84 in *Modern Concepts in Psychiatric Surgery*.

Chapter 15

The human epileptic neuron story can be found in W. H. Calvin, G. A. Ojemann, and A. A. Ward, Jr., "Human cortical neurons in epileptogenic foci: comparison of interictal firing patterns to those of 'epileptic' neurons in animals," *Electroencephalography and Clinical Neurophysiology* 34: 337-351, April 1973. A review of the operant training of epileptic monkeys is A. R. Wyler, E. E. Fetz, and A. A. Ward, Jr., "Firing patterns of epileptic and normal neurons in the chronic alumina focus in undrugged monkeys during different behavioral states," *Brain Research* 98: 1-20, 1975. Anatomical defects in foci have been described by L. E. Westrum, L. E. White, Jr., and A. A. Ward, Jr., "Morphology of the experimental epileptic focus," *Journal of Neurosurgery* 21: 1033-1046, 1964; and in M. E. Scheibel, P. H. Crandall, and A. B. Scheibel, "The hippocampal-dentate complex in temporal lobe epilepsy," *Epilepsia* 15: 55-80, 1974.

1. See W. H. Calvin, "Reexcitation in normal and abnormal repetitive firing of CNS neurons," in *Abnormal Neuronal Discharges*, ed. by N. Chalazonitis and M. Boisson (New York: Raven Press, 1978), pp. 49-61.

Chapter 16

The quotations from Lewis Thomas are reprinted with his permission and that of the *New England Journal of Medicine*, where they originally appeared. They may be found reprinted in the readily available *The Lives of a Cell* (New York: Viking Press, 1974).

Chapter 17

1. The long-term results of operations to remove epileptic foci have been extensively evaluated. About 30-40 percent will be free of all seizures; another 30-40 percent will have one seizure per year or less after an operation, though continuing anticonvulsant drugs. About 10 percent receive no benefit from the operation. See: *Advances in Neurology, vol. 8*, and also R. Rapport, G. Ojemann, A. Wyler, and A. A. Ward, Jr., *Western J. Medicine* 127: 185, 1977.

2. *The numbers game.* It is interesting to ask how much research is being done, relative to the usual industrial expenditures for research and development, which run from 10 percent of the operating budget of a computer or pharmaceutical firm down to a modest 2 percent for a conservative industry such as a utility. The basic and applied research budgets relevant to brain research for the major funding agencies (NIH's National Institute of Neurological and Communicative Disorders and Stroke, the National Institute of Mental Health, and the National Science Foundation) totaled about $400 million in 1979. The 1975 direct and social costs of the diseases under the purview of NINCDS and NIMH totaled at least $100 billion even without including the costs of mental retardation. Thus, the R&D percentage in the "brain business" is perhaps 0.4 percent, and it is unlikely that corrections for the expenditures of pharmaceutical firms and foundations would even double this. So brain-research expenditures seem excessively modest. Some have suggested that an equivalent of the highway trust funds is needed for biomedical research in general—for example, a tax on health-insurance premiums which would go into a research fund.

The sources of the incidence and cost figures are: (1) a 12 March 1979 press release from the National Institute of Neurological and Communicative Disorders and Stroke (NINCDS) in conjunction with Congressional testimony of its Director; (2) the 1977 report of the President's Commission on Mental Health; and (3) J. Bonica, manuscript in preparation. These figures are from the mid-1970s. The $400 million estimate for brain-research expenditures in 1979 was made by totaling the NINCDS budget with the mental-health-research budget of the Alcohol, Drug Abuse, and Mental Health

Administration and the budget of the Behavioral and Neural Sciences division of the National Science Foundation. There is additional related research supported by other NIH institutes (eye, dental, general medical, child development, aging, etc.), the Veteran's Administration, and the pharmaceutical and medical-instrument companies.

There are a number of foundations, usually named after diseases, which also contribute to the overall research budget; the effect of this money, however, is all out of proportion to its dollar amount because it can be utilized more flexibly than government money.

Note on Illustrations

The frontispiece is from S. Ramon y Cajal. The series of birth vs. six years old ink drawings of human cortical neurons is from the work of J. L. Conel, *Postnatal Development of the Human Cerebral Cortex*, vols. 1 (1939) and 8 (1967), reprinted by permission of the Harvard University Press. All ink drawings in that work were made from right brain; in our discussions of them, we have assumed that left-brain areas are identical in structure to the right-brain areas, although that may be untrue in some details. Our labels correspond to the von Economo area designations used by Conel as follows: leg (FA lower extremity), posterior parietal (PH), visual cortex (OC), frontal lobe (FDp), cingulate gyrus (LA), language cortex (PG), hippocampus (HE, HF), arm (PB arm), supplementary motor (FB), auditory (TB), temporal tip (TF).

INDEX

Some technical terms listed here may not appear in the text. The definitions are designed to assist reading reference materials. When page numbers appear after such a definition, an example of the defined condition will be found on the specified page.